Developments in Environmental Modelling, 11

Mathematical Modelling of Environmental and Ecological Systems

Developments in Environmental Modelling

Series Editor: S.E. Jørgensen
                 Langkaer Vaenge 9,
                 3500 Vaerløse,
                 Copenhagen,
                 Denmark

Developments in Environmental Modelling, 11

Mathematical Modelling of Environmental and Ecological Systems

Edited by

Dr. J.B. SHUKLA
Department of Mathematics, Indian Institute of Technology, Kanpur, 208016 India

Dr. T.G. HALLAM
Department of Mathematics and Graduate Program in Ecology, University of Tennessee, Knoxville, TN 37996, U.S.A.

and

Dr. V. CAPASSO
Department of Mathematics, University of Bari, Campus, 70125 Bari, Italy

Excerpts from an International Symposium on 'Mathematical Modelling of Ecological, Environmental and Biological Systems', held from August 27th–30th, 1985 in Kanpur, India.

ELSEVIER
Amsterdam — Oxford — New York — Tokyo   1987

ELSEVIER SCIENCE PUBLISHERS B.V.
Sara Burgerhartstraat 25
P.O. Box 211, 1000 AE Amsterdam, The Netherlands

Distributors for the United States and Canada:

ELSEVIER SCIENCE PUBLISHING COMPANY INC.
52, Vanderbilt Avenue
New York, NY 10017, U.S.A.

Library of Congress Cataloging-in-Publication Data

International Symposium on Mathematical Modelling
   of Ecological, Environmental, and Biological systems
   (1985 : Kanpur, India)
   Mathematical modelling of environmental and ecological
systems.

   (Developments in environmental modelling ; 11)
   1. Ecology--Mathematical models--Congresses.
2. Pollution--Environmental aspects--Mathematical models
--Congresses. I. Shukla, J. B. II. Hallam, T. G.
(Thomas G.) III. Capasso, V. (Vincenzo), 1945-   .
IV. Title. V. Series.
QH541.15.M3I48 1985   574.5'072'4   87-20212
ISBN 0-444-42808-0 (U.S.)

ISBN 0-444-42807-0 (Vol. 11)
ISBN 0-444-41948-9 (Series)

© Elsevier Science Publishers B.V., 1987   *16405667*

Printed in The Netherlands

ACKNOWLEDGEMENTS

It is our privilege to thank the many people who provided moral support, financial assistance, and technical help in connection with the International Symposium on Mathematical Modelling of Ecological, Environmental, and Biological Systems, held in Kanpur, India in August 1985.

We especially note that Ms. Cindi Blair's contribution has been extremely valuable in the preparation of this book. Cindi took time from her many duties in Knoxville to help edit and to type the complete manuscript. Jai Li kindly helped with many phases of the processes from refereeing to proofreading. The assistance of the staff at Elsevier Science Publishers, especially K.C. Plaxton, editor for the Agricultural and Environmental Sciences Section, is gratefully acknowledged.

PREFACE

Problems of developing countries are not always in the forefront of ecological or environmental science. The International Symposium on Mathematical Modelling of Ecological, environmental and Biological Systems was an occasion, perhaps the first one, where scientists from many countries gathered in a developing country for a multidisciplinary forum focussing upon mathematical modelling of ecological and environmental issues. Because of the location of the symposium in Kanpur and because of the large numbers of Indian scientists who participated, it was particularly appropriate that many of the contributions addressed concerns of India.

The spectrum of ecological and environmental topics covered indicated the many disciplines from the engineering, biological, and mathematical sciences represented by those present. The range of papers was from quite theoretical methodology to very practical applications.

This book, an outgrowth of papers presented at the symposium, draws upon the symposium theme and reflects multidisciplinary characteristics of the conference contributions. The choice of topics emphasizes many aspects of ecological and environmental matters including air and water pollution, ecotoxicology, resource management, epidemiology, and population and community ecology.

While abatement of pollution and ecosystem stresses are concerns of the entire world, a lack of attenuating responses is particularly distressful to people in developing countries. It is hoped that this publication will focus international attention upon some problems in the ecological and environmental sciences that can be impacted by mathematical modelling and analysis.

Thomas G. Hallam

FOREWORD

The expansion and complexity of our civilization with the resultant increase in demand for resources has impacted, directly and indirectly, most ecological systems. Though the environment has undergone change since antiquity and the biota has responded to these variations, modern industrial developments have affected some environments to the extent that the future of the biosystems are now nonexistent or, at best, unknown. The quality of the environment, while often relegated to local or national perspectives, is clearly an international concern. Recent international crises such as effects of acid deposition, the radiation release at Chernobyl, and the Sandoz chemical spill in the Rhine River emphasize the drastic impacts of humankind on the biosphere.

As one of the most important problems facing society today is the assessment and abatement of effects of anthropogenic stresses, it is essential to study the problems of ecological and environmental systems from both qualitative and quantitative viewpoints. The attainment of a predictive capacity to determine optimal or adaptive control strategies is necessary to assess risks of exposures. To achieve predictive capability in the biological sciences, mathematical modelling must play an important role in analyzing dynamic system variables using both qualitative and quantitative measures. Recent developments in the mathematical sciences have led to more realistic mathematical models, more informed model analysis, and a better understanding of the relationship between man and his biosphere.

To explore the role of mathematical modelling in analyzing the complex problems of environmental and ecological sciences, an International Symposium on Mathematical Modelling of Ecological, Environmental, and Biological Systems was held at the Indian Institute of Technology, Kanpur during August 27-30, 1985 under the convenorship of Professor J.B. Shukla. Over one hundred delegates including fourteen from foreign countries took part in this international symposium.

A brief chronicle of the opening session indicates the focus of the first conference of this type held in India. The symposium was inaugurated by Profesor R.S. Mishra, a Senior Mathematican and former General President of the Indian Science Congress Association who is currently the Vice-Chancellor of Lucknow University. Professor Mishra commented perceptively upon the need for the symposium. At the inauguration ceremony, Professor T.G. Hallam of the University of Tennessee, Knoxville, U.S.A., in his keynote address, which is reproduced in part in this volume, lamented over problems of environmental pollution that India, in particular, and the world, in general, faces today. Professor S. Sampath, Director of the Indian Institute of Technology, Kanpur in his welcome address pointed to the need of a system science approach to solve these pressing problems so that ecological and environmental structures can be preserved in tune with our needs and aspirations. Professor J.B. Shukla, Convenor of the Symposium, emphasized the necessity of mathematical modelling and especially its role in predicting the highly nonlinear, complex behavior of ecological, environmental, and biological systems. Professor R.N. Biswas, Dean of Research and Development at IIT-Kanpur informed the participants about efforts at IIT-Kanpur in dealing with problems of ecology and environment.

During the four days of the conference, technical sessions on ecosystems, environmental systems, biosystems, and analysis of mathematical models were organized. The papers in this volume represent a cross-section of the papers presented in the areas of environmental and ecological systems. The themes of ecological concern and environmental protection are present in many of the contributions although the primary emphasis is upon understanding and analyzing the mathematical models and their implications for solving the scientific problems.

General environmental areas addressed in this book include mathematical models for air pollution, water pollution, and ecotoxicology. Ecological developments focus on the applied areas of resource management and epidemiology as well as some theoretical aspects of population and community ecology.

T.G. Hallam
J.B. Shukla
V. Capasso

LIST OF AUTHORS

E. BERETTA - Instituto di Biomatematica, Universita di Urbino - Via Saffi, 1
61029 Urbino, ITALY

V. CAPASSO - Dipartimento di Matematica, Universita di Bari - Campus 70125 Bari,
ITALY

R.S. CHAUHAN - Department of Mathematics, Indian Institute of Technology,
Kanpur, 208016 INDIA

A. CIGNO - Department of Economics and Commerce, University of Hull,
Hull, HU6 7RX ENGLAND

J.M. CUSHING - Department of Mathematics, University of Arizona, Building 89,
Tucson, Arizona 85721 U.S.A.

H.I. FREEDMAN - Department of Mathematics, University of Alberta,
Edmonton, Alberta CANADA

V. GUPTA - Department of Mathematics, Harcourt Butler Technological Institute,
Kanpur, 208016 INDIA

V.K. GUPTA - Safety Review Committee, Department of Atomic Energy, Bhabha
Atomic Research Centre, Trombay, Bombay, 400085 INDIA

T.G. HALLAM - Department of Mathematics and Graduate Program in Ecology
University of Tennessee, Knoxville, Tennessee 37996-1300 U.S.A.

R.K. KAPOOR - Safety Review Committee, Department of Atomic Energy,
Bhabha Atomic Research Centre, Trombay, Bombay, 400085 INDIA

Y. KITABATAKE - National Institute for Environmental Studies, Vatabe-machi
Tsukuba, Ibaraki 305 JAPAN

L.K. PETERS - Department of Chemical Engineering, University of Kentucky
Lexington, Kentucky 40506-0046 U.S.A.

B. RAI - Department of Mathematics, University of Allahabad, Allahabad, INDIA

M. SALEEM - Department of Mathematics, Harcourt Butler Technical Institute
Kanpur, 208016 INDIA

A. SHUKLA - National Technical Manpower Information System (Nodal Centre)
Indian Institute of Technology, Kanpur, 208016 INDIA

J.B. SHUKLA - Department of Mathematics, Indian Institute of Technology
Kanpur, 208016 INDIA

V.P. SHUKLA - Computer Division, CW & PRS, Pune, 411024 INDIA

S.U. SIDDIQUI - Department of Mathematics, Harcourt Butler Technical Institute
Kanpur 208016 INDIA

R. SMITH - Department of Applied Mathematics and Theoretical Physics, University of
Cambridge, Silver Street ,Cambridge CB3 9EW ENGLAND

TABLE OF CONTENTS

KEYNOTE ADDRESS

Some Environmental Concerns and Flickering Prospects of Relief*
    by Thomas G. Hallam

SOME ENVIRONMENTAL CONCERNS AND FLICKERING PROSPECTS OF RELIEF*

THOMAS G. HALLAM
Department of Mathematics and Graduate Program in Ecology, University of Tennessee, Knoxville, Tennessee

Distinguished Scientists and Honored Guests:

First, I would like to express my appreciation to Professor Shukla, Convenor of the Symposium, and the organizing committee for all their work. It is a big job to organize a conference the size of this one; their efforts are deeply appreciated.

You are probably wondering why I, a lowly professor, am your keynote speaker. There are several good reasons. First, Prime Minister Rajiv Gandhi did not come - but as you can note in the Souvenir Program, he sent a letter of welcome. Secondly, the state minister of U.P. could not come although he was scheduled to speak. Thirdly, I was the only visiting participant here early last week.

I do not claim expertise in the ecology nor the environment of INDIA - however, judging from certain politicians' actions, expertice does not appear to be a qualification for speaking on a subject - so I shall proceed. I begin with an apology as I plan to relate to you some personal impressions - an uninformed ideosyncratic viewpoint - on what many of you already know about India.

Before arriving in India, because of what I had read, because of what I had heard, I had a grave concern about the awareness by Indians in general about the ecology of India and the deterioriating environment in India. All signs indicated there was a lack of interest in ecological and environmental issues in this country. My impression was that this perspective was placable - one's main concern here was day-to-day maintenance and survival.

Just the existence of this conference, with the large total number of local participants, shows awareness, interest and concern are present here. There are many other positive signs I have seen as well and I will return soon to some of them.

After being here for two weeks, my concern about ecological awareness has weakened and I will tell you why shortly. But my concern for the condition of the Indian environment has not and I will tell you why shortly.

I shall relate some personal examples and recent experiences as well as relay some information obtained locally. My primary scientific sources include the reputable Pioneer of Lucknow and the Hindustan Times of Delhi. Reading local English language newspapers has been a pleasant surprise - there is much concern about ecology and problems of the environment.

---

*Excerpts from the keynote address delivered at the International Symposium on Mathematical Modelling of Ecological, Environmental, and Biological Systems, Kanpur, U.P. India, August 1985. The viewpoints expressed are those of the author who is solely responsible for their content.

ENVIRONMENTAL PROBLEMS. The first local paper I picked up after arriving in India contained an article about a professor in Bombay who was working in an area of water pollution. He experienced a difficulty many of us have faced - funding had just been eliminated from his research project. The reason given for this lack of support was "too much poverty in India". While there may be valid reasons for not funding this particular proposal, the reason stated is certainly not one. The poor are the ones most affected by environmental problems.

During my stay here, I visited the Taj Mahal. It is a fantastic place - the white marble structure is surely one of the most spectacular physical representations of love on this Earth. It is approximately 400 years old. It has survived wars and storms yet it looks almost new. Not for long!! A refinery has been built in the area. Emissions are already creating problems of deterioration.

GANGA RIVER. One of the most significant geographical features in this area is the famous GANGA river or as we English language people say, the Ganges. I have noticed many recent articles on the Ganges. Prime MInister Rajiv Gandhi's Independence Day address mentioned the Ganga River as an area of concern and emphasized the need to clean it up. An article discussing the pollution of the Ganga indicated the water is too dirty and dangerous for a bath - religious or otherwise. One city on the Ganga has no sewage treatment plant, but it has sixty-five open drains into the river. Bodies, cremated, partially cremated, and not partially cremated, are dumped into the river as a holy burial place. Industrial pollution is rampant. One area of the river ignited and burned for several days - while this is not unique to the Ganges, it is certainly an indication of the magnitude of the problem. In Kanpur, many effluents come from local tanneries. The city of this conference is cited as one of the two worst polluters in Uttar Predesh, the biggest state in India. Kanpur's and many other cities' drinking water supply comes from the Ganga.

CHEMICAL POLLUTION. Some chemical companies are allowing the destruction of wildlife and even citizens of India as well as those of other countries. The recent Bhopal tragedy lies heavily on our hearts. The chemical DDT, still in use here after widespread bans elsewhere, is inexpensive to use now but, in the long run, it is disasterous especially for wildlfe - primarily because of food chain accumulation and persistence. It effects humans directly and indirectly (by destroying large birds that eat rodents which are vectors for diseases). Newspapers indicate that one hundred thousand tons of pesticides are used per year in India. Of this quantity, seventy percent are banned or on restricted lists in western countries.

DISEASES. Firewood women so painstakingly collect is burnt in highly smoky inefficient "chulhas". The wood smoke eminating from a chulha contains forty times the

World Health Organization recommended safe level of suspended particulates. In three hours exposure, they inhale carcinogenic Benz(A) pyrene amounts equal to two hundred cigarettes.

Resistence to pesticides has evolved to a state where mosquito-borne diseases such as malaria and encephalitus are increasing.

It is a sad situation when the government is among the biggest polluters. The areas of electrical power generation, sewage disposal and proper drainage are distressful.

What I have done is to list some crucial problems I have seen in a short time. I read of others - 1.5 million acres of forest cut per year - soil erosion - recharge of ground water reduced. Don't feel I am singling out India - one can always find negative things anywhere. Indeed, in my opinion, the United States has considerably decreased its responsibility to the environment in recent years. But, I think important thresholds are now being reached in some countries and time for damage repair is running out.

ANTICIPATION AND ASPIRATIONS. I know, this week will be exciting because, without exceptions, each of these areas of concern is being addressed at this conference as are many other problems.

One cannot always find positive occurrences to emphasize. Since my arrival here I have seen some very assertive measures of ecological progress. These include the fact that the Bengal tigers should survive. In 1974 there were approximately 1200 tigers; today, there are approximately 4000 including more than 1200 in protected environments. The article also indicated there was a good sex ratio - whatever that means.

I had the priviledge to visit a crocodile farm outside of Lucknow. Not long ago it was estimated that not more than sixty gharial crocodiles survived in the wild. There is a 99% mortality from the egg stage to a juvenile of three years. The farm has achieved a surprising 60% survival rate until release at age three.

I also visited a small university laboratory where fly ash is being treated chemically with success.

A CHALLENGE. We are attending this International Symposium on Mathematical Modelling of Ecological, Environmental, and Biological Systems with an awareness of the problems that exist here and throughout the world. We are cognizant of the progress that is being made but there are still many concerns and questions.

If I may address the scientists from India. In your country environmental laws exist - enforcement does not. In your country, ecological protection laws exist - should they be enforced at the expense of poverty? In your country, resources are not well managed - it is estimated that 2.6 billion people can be supported here by a well disciplined agricultural system but the country is struggling to feed its citizens.

Problems like the Ganga River with its immense length look impossible. In mathematics we have impossible and possible problems. We work on the possible ones. The Ganga is a possible problem - at least for a while until it, as a complete deteriorated environment, dies. There are numerous success stories- the Thames and Rhine Rivers, Lake Erie - for polluted aquatic systems. But the Ganga is an Indian problem; it will require an Indian solution. Isn't a religious and physical lifeline of India worth it? [Since this talk was given, a large scale clean-up of the Ganga has begun.]

These problems are certainly difficult and challenging. Aspects of your educational system are hinderances. How can one attack biological problems where it is necessary to make predictions about the future structure of an ecosystem without biologists being able to talk to engineers and mathematicians or vice versa? At IIT-Kanpur there is no biology department. Students in environmental mathematics often take only mathematics. It is difficult to attack multidisciplinary problems when your knowledge is only disciplinary.

Aspects of your culture are a hinderance. In spite of your democracy and reverence for equality, not all things are equal. A person has uncontested rights as long as they do not interfere with the rights of others. When populations get large, density dependent effects come into play. Couples with many children are affecting the rights of others - something must be done with your number one problem - overpopulation.

In spite of many difficulties, there is no rational reason for toleration of environmental conditions as they are. Poverty is no excuse - much can be done at low cost; however; some things must be done independent of cost. Ignorance is no excuse - much environmental and ecological information exists in India and in the world. Manpower is no excuse - you have many highly trained people and an overabundance of others. Use them all!! A lack of a solution is no excuse - one must be found for many problems.

CONCLUDING REMARKS. We are gathered here to talk about mathematical modelling in environmental and ecological problems and to see the current state of the art. In my opinion, that is not enough of an objective for this conference. My hope is not only for a free exchange of ideas but a renewed dedication by each Indian scientist present to commit yourself, if you have not done so, to work actively to improve your environment. If each of you makes this commitment there will be an additional national nucleus for the improvement of the ecological structure of your country. With such a force, you can expect and will receive help from scientists from around the world. If you do not make such a commitment, you, citizens of a country where 99% of the people would not hurt any living thing, because of inaction are hurting yourself, your family, and country.

It is time for action. Take time to form a conference political action group. Contact your m.p. and state conservation offices. Get them to respond to your needs and support them as well. Volunteer to help and, most importantly, to lead. Professors and students have much influence and command much respect in India - use it. If you don't know what to do now that most of you have changed to a five-day work week, use the sixth day for the environment.

Take pride in your heritage - renew it. Remember you are not alone in struggles like these. Let this conference be strong, truthful and productive. Let us discuss and debate the real issues as interested devoted scientists.

PART I - AIR POLLUTION

SOME ASPECTS OF MATHEMATICAL MODELING OF ATMOSPHERIC TRANSPORT AND CHEMISTRY

LEONARD K. PETERS
Department of Chemical Engineering, University of Kentucky, Lexington, Kentucky 40506-0046

1 ABSTRACT

Mathematical modeling of trace species in the atmosphere can help the scientist better understand the complex processes occurring. Many of the relevant species (e.g., SO_2, sulfate, NO_x, and CO) have atmospheric lifetimes that suggest regional and global-scale physico-chemical processes control their behavior. The interactions among the transport, chemistry, and removal processes can be simulated by numerical solution of the conservation of mass equations for the respective species. The important meteorological factors affecting their distributions include the wind field, mixing layer dynamics, and solar radiation intensity.

In this paper, an overview of regional-scale atmospheric transport/chemistry modeling is presented. The individual atmospheric processes affecting species distributions are quite complex, so that parameterizations are frequently necessary for an integrated, complete mathematical model analysis. The equations to be solved are summarized, some parameterizations are presented, and specific removal mechanisms are discussed. Simulation results from a three-dimensional, time varying regional SO_2 and sulfate atmospheric model are discussed. Substantial spatial and temporal variations in the distributions of SO_2 and sulfate are predicted.

2 INTRODUCTION

Many complex chemical and physical processes that determine the concentrations of trace species in the atmosphere occur simultaneously. Field experiments in which most variables are controlled are generally not possible. As a result, the atmospheric scientist frequently is required to combine data from controlled laboratory experiments with theoretical analysis to construct comprehensive mathematical models of the interacting phenomena. These models can enhance our understanding of several important points: (1) which species are present and at what concentrations and with what distribution (spatial, temporal, geographic); (2) which processes control the cycles of trace species; (3) how do these species interact with climatic factors; and (4) how do humans impact on the cycles (both now and in the future) (Peters and Carmichael (1983)).

In this paper, an overview of atmospheric transport and chemistry models will be presented. Particular emphasis will be placed on regional models, where the horizontal length scales are on the order of hundreds of kilometers and time scales are on the order of days. The structure of these tropospheric models is inherently process oriented. The transport, chemistry, dry deposition (surface removal), wet removal (cloud and precipitation processes), and primary source processes occurring in the troposphere determine the residence time and cycles of trace species. Thus, to develop meaningful models, it is necessary to characterize mathematically the chemical and physical processes. Each process treated in the model should be treated in appropriate detail. For example, if in a particular application the transport and chemistry are of comparable importance, the model should treat these processes in similar detail. A very detailed chemistry model with little or no transport would be of marginal value, as would a transport model with no chemistry.

Each individual atmospheric process is itself a very complex phenomenon. Hopefully, one does not have to wait until the details of the phenomena are completely understood to simulate the process, but simulation can be accomplished by utilizing suitable chemical, dynamic, and thermodynamic parameterizations. At the present, it seems possible to develop realistic models that include the governing processes and still consider a large number of chemical species.

3 MATHEMATICAL BASIS

3.1 Governing Equations

The turbulent transport of material in the atmosphere can be analyzed from either an Eulerian or a Lagrangian approach. The differences between these approaches involve the way the position in the field is identified. In the Eulerian approach, the concentration of material is described at a given point and time with the coordinates fixed and independent of time. Therefore, the coordinate position $\mathbf{x}$ (e.g., in Cartesian coordinate system $\mathbf{x} = (x, y, z)$) and time are independent variables.

In the Lagrangian approach, the concentration of material is described for a particular fluid element as it travels with the flow. The coordinate position $\mathbf{x}$ is that of the fluid element and, in the general case, depends on time. Therefore, in the Lagrangian approach the coordinates are dependent variables, and the fluid element is identified by its position in the field relative to that at some arbitrary time, usually $t = 0$.

In this paper, we will emphasize Eulerian models. The mathematical equation describing the time-averaged concentration of a reactive species using the Eulerian approach is

$$\frac{\partial \langle c_1 \rangle}{\partial t} + \nabla \cdot (\langle \mathbf{v} \rangle \langle c_i \rangle) = -\nabla \cdot (\langle \mathbf{v'} c_i' \rangle) + \langle S_i \rangle + \langle R_i \rangle + \langle G_i \rangle, \tag{1}$$

where $\langle \mathbf{v} \rangle$ is the velocity vector, $\langle c_i \rangle$ is the concentration of species i, $\langle S_i \rangle$ is the source term, $\langle R_i \rangle$ is the formation rate by chemical reaction, and $\langle G_i \rangle$ accounts for transfer from the gas phase to an aqueous phase in the form of clouds, precipitation, etc. The $\langle - \rangle$ represents a time-averaged quantity.

The first term on the right-hand side of Equation (1) is introduced by the time-averaging procedure and represents the turbulent fluxes. This term introduces two new variables, $\mathbf{v'}$ and c_i', the turbulent fluctuations of the velocity and concentration, respectively. Therefore, to solve Equation 1, it is also necessary to have equations for $\langle \mathbf{v'}c_i' \rangle$. Such equations can be derived, but they ultimately depend on terms such as $\langle v_j' v_k' c_i' \rangle$. Each attempt to derive an expression for the terms introduced by the time-averaging process introduces new, higher-order terms. Thus, there is no rigorous way of "closing" the system.

The most common method of closure for Equation (1) is to model the turbulent fluxes analogously to Fickian diffusion, i.e., as a product of a diffusion coefficient and the mean concentration gradient,

$$\langle -\mathbf{v'}c_i' \rangle = \mathbf{K} \cdot \nabla \langle c_i \rangle , \tag{2}$$

where $\mathbf{K}$, the eddy diffusivity tensor, is a function of the fluid flow but not of the material property (Seinfeld (1975)). This closure is referred to as K-theory. Higher-order closure models are being developed, but at present they are computationally expensive (Donaldson and Hilst (1972); Lewellen and Teske (1975, 1976)), and some results indicate that their effect is not substantial (Kewley (1978)).

The mathematical basis of Lagrangian transport of a nonreactive material is

$$\langle c(\mathbf{x}, t) \rangle = \int_0^t \int_{-\infty}^{\infty} P(\mathbf{x}, t|\mathbf{x'}, t')S(\mathbf{x'}, t')d\mathbf{x'}dt' , \tag{3}$$

where $P(\mathbf{x}, t|\mathbf{x'}, t')$ is the probability density that a particle released at $\mathbf{x'}$ at time t' will be found at point $\mathbf{x}$ at time t, and S is the function describing the source distribution. The Lagrangian approach is free from the closure problem associated with the Eulerian approach. However, the solution of Equation (3) requires that the probability density function be specified. It is interesting to note that if the process is assumed to be Markovian (i.e., the process is stochastic and the future state depends only on the present state and on the transition probabilities from the present to the future state), Equation (3) reduces to

$$\frac{\partial \langle c \rangle}{\partial t} + \nabla \cdot (\langle \mathbf{v} \rangle \langle c \rangle) = \nabla \cdot \mathbf{K} \cdot \nabla \langle c \rangle + \langle S \rangle , \tag{4}$$

which is the same equation arising from the Eulerian approach with use of K-theory closure for an inert species.

Many regional-scale transport models currently in use are a form of Lagrangian model frequently referred to as trajectory models (Shannon (1981)). These models are generally formulated under assumptions of no horizontal turbulent diffusion, no convergent or divergent flows, and no wind shear. In these models parcels of material are emitted from each source and are advected with the mean wind with the parcel's location computed at equal time intervals. By following a fixed mass of air, trajectory models avoid lengthy integration of Equation (4) and permit the general verification of pollutant distribution by tracing parcels from its source area.

Eulerian models employ numerical integration of Equation (1) and provide solutions over the entire modeling region, rather than along trajectories. These grid models are also able to include the effects of wind shear and vertical air motions.

When the material of interest is chemically reactive, $\langle R_i \rangle$ is frequently evaluated by using the mean concentrations. However, it must be emphasized that in obtaining Equation (1), the time averaging of $\langle R_i \rangle$ may introduce additional terms; e.g., if

$\langle R_i \rangle = -k\langle c_i \rangle^2$, then $R_i = -k(\langle c_i \rangle + \langle c_i'^2 \rangle)$. So again, a closure problem arises when

turbulent flow is considered. The influence of turbulence on the reaction rate is most significant in near-source areas. In addition to these considerations, the reaction term can greatly complicate the system because $\langle R_i \rangle$ can be highly nonlinear and can, therefore, introduce complicated coupling of the equations.

In general, tropospheric trace distributions exhibit large variations in all three dimensions. Therefore, simplification of models to one or two dimensions severely restricts the possibility of making meaningful comparisons between model calculations and measured data. This is particularly true for those compounds that have a residence time in the atmosphere shorter than the characteristic transport time. Thus, modeling for such species requires three-dimensional models. Both regional (for SO_x compounds) and global (CH_4 - CO) three-dimensional combined transport-chemistry models have been developed (Carmichael and Peters (1980, 1984a,b); Peters and Jouvanis (1979)). In addition, detailed three-dimensional global transport models with limited chemistry have been developed (e.g., Mahlman, Levy, and Moxim (1980)).

3.2 Boundary Conditions

In addition to the governing equations [Equation (1)], the boundary conditions must be satisfied. At the surface, the boundary condition typically takes the form

$$K_z \frac{\partial \langle c_i \rangle}{\partial z} = v_{di} \langle c_i \rangle - \langle Q_i \rangle, \text{ line} \qquad (5)$$

where v_{di} is the deposition velocity, and $\langle Q_i \rangle$ is the surface emission flux. The term deposition velocity apparently was introduced in the micrometeorology community, so named because of the units on v_{di}. It represents an empirical exchange coefficient to describe material transfer from the air to the underlying surface (i.e., a mass transfer coefficient in engineering parlance).

At the other boundaries, the condition is that for the prescribed total flux; i.e.,

$$\langle v \rangle \langle c_i \rangle - K \cdot \nabla \langle c_i \rangle = F_i \ . \tag{6}$$

The form of the prescribed flux F_i will depend on whether there is inflow or outflow at that boundary. The most typically used conditions are a zero diffusive flux (i.e., $-K \cdot \nabla \langle c_i \rangle = 0$) under outflow, and the total flux equal to the advection flux under inflow conditions (i.e., $F = vc_{bi}$, where c_{bi} represents a background concentration). However, these conditions are restricted to applications where sources and sinks are located far from boundaries and reaction rates are sufficiently fast that much of the conversion has already occurred by the time the air mass reaches the boundary. Alternative conditions have been proposed (Carmichael and Peters (1980)).

3.3 Numerics

In general, closed-form analytic solutions to Equation (1) do not exist for the initial and boundary conditions encountered in complex situations. Therefore, the solutions frequently must be obtained numerically. However, numerical simulation of this equation is difficult for several reasons: (1) the problem is multi-dimensional; (2) the horizontal transport is usually convection dominated; (3) the boundary conditions are mixed; and (4) both slow and fast chemical reactions can be important.

There are many methods available for solving atmospheric transport problems. The methods available can be broadly classified into the categories of finite-difference schemes, variational methods, and particle-in-cell techniques. The method we are presently using is a combination of the concept of fractional time steps and one-dimensional finite elements. This is referred to as the Locally One-Dimensional, Finite-Element Method (LOD-FEM). The LOD procedures (Mitchell (1969)) split the multi-dimensional partial differential equation into time dependent, one-dimensional problems which are solved sequentially. The time-split equations are of the form

$$\frac{\partial \langle c_i \rangle}{\partial t} + L_x \langle c_i \rangle = 0 \ , \tag{7}$$

$$\frac{\partial \langle c_i \rangle}{\partial t} + L_y \langle c_i \rangle = 0 \ , \tag{8}$$

$$\frac{\partial \langle c_i \rangle}{\partial t} + L_z \langle c_i \rangle = 0 \ , \tag{9}$$

$$\frac{\partial \langle c_i \rangle}{\partial t} = \langle R_i \rangle - \langle G_i \rangle \ . \tag{10}$$

L_x, L_y, and L_z represent the one-dimensional transport operators, e.g.,

$$L_x \langle c_i \rangle = \frac{\partial (\langle u \rangle \langle c_i \rangle)}{\partial x} - \frac{\partial}{\partial x} (K_{xx} \frac{\partial \langle c_i \rangle}{\partial x}) .$$

Equations for the cloud and precipitation drop phases are of similar form. The chemical reaction and mass transfer terms are treated separately because these can have time scales much smaller than that for the transport. Splitting out these terms allows different time steps to be specified for the transport and for the chemistry/removal processes.

The transport equations are solved using a Crank-Nicolson Galerkin finite element technique. An upwinding parameter (Heinrich, Huyakorn, Zienkiewicz, and Mitchell (1977)) and spatial filtering, as originally proposed by Forester (1977) and subsequently described by McRae, Goodin, and Seinfeld (1982), can be included. An important feature of this numerical procedure is that the computer code can be structured so that any combination of Equations (7) through (10) can be used, enabling application to one, two, or three-dimensional time-dependent analysis.

The technique used to integrate the chemistry and mass transfer terms is an adaptation of the semi-implicit Euler method proposed by Preussner and Brand (1981). This method consists of writing Equation (10) in the form

$$\frac{\partial \langle c_i \rangle}{\partial t} = -\langle G_i \rangle , \tag{11}$$

$$\frac{\partial \langle c_i \rangle}{\partial t} = \langle R_i \rangle = \langle P_i \rangle - \langle D_i \rangle \langle c_i \rangle . \tag{12}$$

$\langle P_i \rangle$ and $\langle D_i \rangle$ represent the chemical production and destruction terms. Over the time step $\Delta t_{j_{chem}}$, Equations (11) and (12) are solved analytically, assuming all other concentrations are constant during the time period. For cases when $\langle D_i \rangle$ is small (e.g., HNO_3 at night), Equation (12) is solved by Taylor Series expansion. This procedure has the important property that $\langle c_i \rangle$ can never become negative, provided that the reaction rate constants are positive and the initial values are non-negative. Further-more, in the limit as $t \to \infty$, the proper steady state concentrations are obtained; i.e.,

$$\langle c_i^{ss} \rangle = \frac{\langle P_i \rangle}{\langle D_i \rangle} . \tag{13}$$

Equation (12) is used to calculate advected species concentrations only. However, $\langle P_i \rangle$ and $\langle D_i \rangle$ also contain concentrations of the short-lived gas-phase species (e.g., free radicals like OH and HO_2) , which are calculated using appropriate pseudo-steady state approximations. This procedure results in algebraic equations that depend on the concentrations of the advected species. The calculation procedure to advance from t_j to $t_j + \Delta t_{j_{chem}}$ uses $c_i|_{t_j}$ to calculate the short-lived species concentrations, and then uses these concentrations to advance to $c_i|_{t_j + \Delta t_{j_{chem}}}$ using Equation (12).

The accuracy of this technique depends on the species of interest and $\Delta t_{j_{chem}}$ For relatively long-lived species (e.g., SO_2, CO, and alkanes), relatively large $\Delta t_{j_{chem}}$ can be used, but for more reactive species much smaller time steps are required. This pseudo-analytic technique can execute twice as fast as the Gear method for chemical schemes that we have used.

The LOD-FEM procedure has been evaluated by several workers including the present author [cf. Carmichael, Kitada, and Peters (1980) and Kitada, Carmichael, and Peters (1983)]. More recently, Kasibhatla (1984) has evaluated the advection algorithm used by investigating the accuracy of the solution of the one-dimensional advection equation.

$$\frac{\partial \langle c \rangle}{\partial t} + \langle u \rangle \frac{\partial \langle c \rangle}{\partial x} = 0 \tag{14}$$

Advection of ideal waveforms with a constant wind has been a traditional approach. Kasibhatla's tests were based on advection of triangle, sine, and square wave distributions of moderately narrow width. The triangle and square wave tests have very sharp gradients with discontinuous first derivatives at isolated position(s) in the waveform.

Selected results from Kasibhatla's study are shown in Figure 1 for the three waveforms. These results demonstrate that the Crank-Nicolson Gakerkin method, when applied to an advection-dominated problem, gives rise to numerical oscillations, primarily in the wake of the advected waveform. However, the upwinding scheme substantially minimizes this problem. Part of the diffusion that occurs results from filtering of the data. The reduction in peak value of the concentration depends on the shape of the waveform. These results also clearly show the advantages of using the upwinding scheme and the filtering routine. Greater stability of the numerical solution is obtained and the method is readily applied to high Peclet number ($u \Delta x / K_{xx}$) solutions provided the Courant number ($u \Delta t / \Delta x$) is less than 1. Similar results for multi-dimensional simulations of advection/diffusion have been found also.

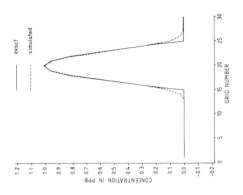

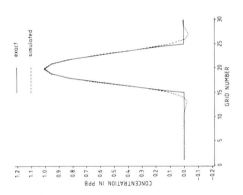

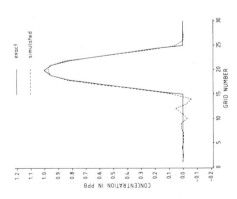

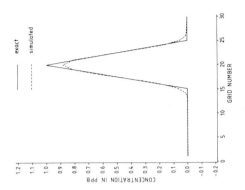

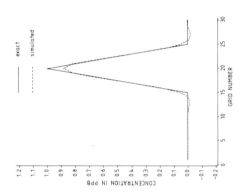

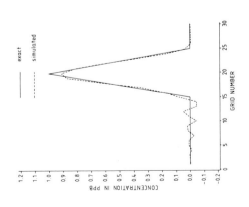

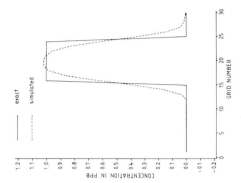

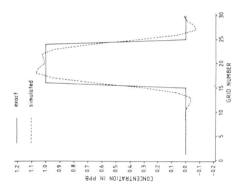

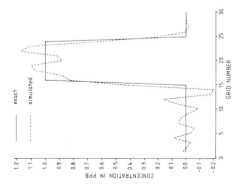

FIGURE 1   One-Dimensional Pure Advection Test of the Crank-Nicolson Galerkin Finite
Element Method with Piecewise Linear Basis Functions. Sine Wave,
Triangle Wave, and Square Wave initial wave forms are shown. Plotted are
results after 24 hours of transport (~ 96 time steps) for the cases without
upwinding and without filter (left column), with upwinding and without filter
(middle), and with upwinding and filter (right column) .

4 CHEMICAL PROCESSES

4.1 General

The numerous trace gas concentrations present in the atmosphere occur from both natural and anthropogenic sources. These trace gases can interact chemically with one another, and since there are many such species, it is difficult (and almost impossible) to discuss one species without considering one or more others. Gaseous compounds show varying degrees of chemical reactivity and, as a result, may exist for very short to very long times in the atmosphere. Figure 2 shows many of the important trace gas species (Peters (1983)). This figure has divided the complex interactive chemistry into several subsets. These subdivisions are somewhat artificial, but it is hoped that it will help the presentation.

Those compounds in the four blocks are stable molecular species and can exist under some conditions for rather long periods of time. Nevertheless, it must be emphasized that even with these stable compounds, there is a wide range of atmospheric lifetimes. For example, NO and NO_2 may only exist for a few hours to a few days, whereas the residence time of CH_4 is of the order of 4-7 years and that for CO_2 is probably around a decade or longer.

The species in the central circle are on the other end of the residence time spectrum. These are the very reactive free radicals that are transient and have residence times that can be measured in seconds, milliseconds, and even shorter times. The subsets contained in the blocks interact through these free radicals contained in the circle. The species listed on the lines, which connect the species subset blocks with the free radical circle, generally have atmospheric residence times intermediate to the free radicals and the more stable molecular species. They are generally produced in situ and do not have significant, identifiable natural or anthropogenic sources at ground level.

4.2 Photochemical Initiation

In the troposphere, the temperature varies from about 215 K to about 315 K, a relatively narrow range of temperatures compared, for example, to the temperature ranges for chemical reactions in combustion processes. Over this rather limited range, chemical reaction rates will generally exhibit relatively small change. These changes may be only over one or two orders of magnitude, unlike many other applications of scientific interest. In addition, many important gas-phase chemical processes occurring in the troposphere do not involve catalysts.

13

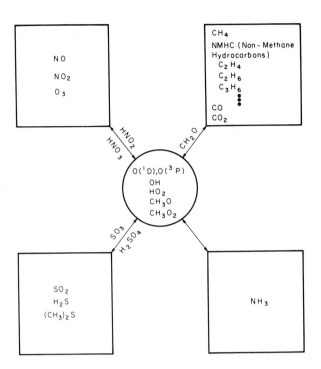

FIGURE 2 Separation of Important Trace Gas Species into Subsets Illustrating the Interactions through the Free Radicals. (From Peters (1983)).

As a result of both of the above restrictions, the energetics for breaking chemical bonds and initiating chemical change in the atmosphere are frequently derived from the sun through photochemical reactions. Notable examples are NO_2 and O_3. In addition, many chemical reactions of atmospheric importance involve free radical species. These free radicals are generally very reactive and exist at quite low concentrations. Under such conditions, their rates of formation and destruction by chemical reactions are nearly in balance. The destruction rate generally depends on the free radical concentration in a linear fashion.

The formation and destruction pathways for the radicals rapidly adjust so that the formation and destruction rates are in close balance. Thus, the pseudo-steady state approximation [i.e., Equation (13)] is said to apply. The pseudo-steady state approximation, if carefully applied, can alleviate mathematical stiffness problems and can reduce the number of species that must be described by differential equations.

4.3 SO_2 Chemistry

In the space available, only limited discussion of a specific species is possible. We will make a few comments about the environmentally important species SO_2.

Current research seems conclusive in establishing that sulfate in the atmosphere primarily comes from the oxidation of SO_2. Both homogeneous and heterogeneous chemical reactions contribute to the conversion of SO_2 to sulfate, although our present knowledge of the heterogeneous processes (e.g., on particles and in condensed cloud water) is limited.

The conversion of SO_2 to sulfate in the atmosphere is quite variable and shows substantial differences between day and night. Through the day, photochemical processes ultimately leading to reactive free radicals are judged to dominate. The hydroxyl and hydroperoxy (OH and HO_2) radicals are the most important in the oxication of SO_2.

$$SO_2 + OH \rightarrow HSO_3 \tag{15}$$

$$SO_2 + HO_2 \rightarrow SO_3 + OH \tag{16}$$

Reaction (15) apparently dominates; and the fate of the HSO_3 radical is not fully understood, although it is generally assumed that sulfate is the ultimate product. Reaction with O_2 to form HSO_5 has been suggested as the next step in the oxidation process. However, a recent model study by Stockwell and Calvert (1983) indicates that a more suitable fate for the HSO_5 free radical may be decomposition into SO_3 and HO_2. This suggestion is important from two aspects. First, SO_3 is formed directly; and secondly, the reaction is not chain terminating, since HO_2 is produced and can participate in other reactions. However, simulations representative of rather polluted air mass concentration levels using this overall scheme have not shown substantial differences from that described by Reaction (15).

Reaction (16) is quite efficient since it has not only oxidized an SO_2 molecule but has also generated another free radical, which can oxidize another molecule of SO_2 or other species. The SO_3 that is produced is short-lived since it can rapidly combine with water vapor,

$$SO_3 + H_2O \rightarrow H_2SO_4 , \tag{17}$$

and nucleate to form sulfate aerosol.

During the daytime, SO_2 conversion rates in clear air conditions are probably around $0.5 - 3\%$ h^{-1}. In-cloud processes can substantially increase the conversion rate. Hegg and Hobbs (1981) have reported rates as high as $4 - 300\%$ h^{-1} (i.e., $0.067 - 5\%$ min^{-1}) under such conditions. During the night, heterogeneous processes are believed to dominate. These may be non-catalytic or catalytic, and may occur on solid particles or liquid (or liquid covered) particles.

Some aspects of chemical models can be compared against carefully controlled field experiments. For example, the homogeneous gas-phase chemistry of SO_2 has been studied by evaluating the conversion rate of SO_2 to sulfate. These results have shown reasonable agreement with some field experiments. Balko and Peters (1983) modeled the August, 1975 field experiments conducted within the urban St. Louis plume (Alkezweeny and Powell (1977)). The St. Louis plume is characterized by relatively high hydrocarbon/NO_x ratios and substantial SO_2 concentrations. Figure 3 shows the ratio of SO_2 to total sulfur as a function of time for both observed and simulated values. In both cases, surprisingly high conversion rates $(10\text{-}14\%\ h^{-1})$ were reported. This shows that, in the presence of high hydrocarbon/NO_x ratios, high gas-phase conversion rates of SO_2 can be predicted, agreeing reasonably well with some field experiments.

5 MODEL INPUTS

Inputs necessary to solve Equation (1) can be classified broadly into emission, meteorological, reaction, and surface removal data. The emission input data include the specification of the emission rates of each transported species over the entire spatial and temporal range of interest (i.e., **x** and t) . The necessary meteorological inputs consist of the transport variables, **v** and **K** , and the reaction rate variables of temperature, water vapor content, solar actinic flux, and cloud cover. The

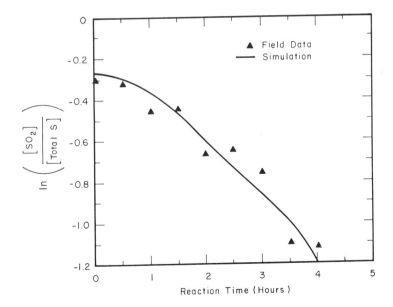

FIGURE 3 Ratio of SO_2 to Total Sulfur for the St. Louis Plume, August 10, 1975. (From Balko and Peters (1983)) .

meteorological variables must also be specified over the entire domain. Inputs classified as reaction and surface removal include the chemical reaction mechanism, rate expressions and rate constants, and dry deposition velocities (which necessitate input of surface type and land-use data).

The required input data can be obtained from observational data and/or can be generated from primitive equation models (e.g., general circulation models). If observational data are used, the problem arises that observations are made only at discrete locations and times. Therefore, temporal and spatial interpolation are required for generation of the input fields. In addition, many of the important input data are seldom measured, for example, the vertical velocity component and K. These quantities must then be estimated from those variables that are measured. Thus, there are large uncertainties in input fields generated from observational data.

The alternative to using observational data directly is to generate input fields by use of a primitive equation model. For global-scale inputs, a general circulation model can be used (Mahlman, Levy, and Moxim (1980)), whereas for regional applications a mesoscale model is more applicable (Tarbell, Warner, and Anthes (1981)). These models, which involve solving the Navier-Stokes equations as well as equations for the energy balance and the water vapor balance, provide as outputs temporally varying three-dimensional meteorological fields. Furthermore, these models can in principle be constrained with the observational fields.

The generation of the meteorological fields from primitive equation models constrained with observational data may improve some of the input fields. However, there still remain large uncertainties. Some of the uncertainties in the input parameters are summarized in Table 1. One of these is the wind field. It is generally recognized that mass-conservative wind fields are required as input to Eulerian models to avoid introducing systematic numerical errors. Several methods have been suggested to generate mass-consistent wind fields, ranging from direct differencing procedures, to variational methods, to iterative interpolation methods (Goodin, McRae, and Seinfeld (1980); Liu and Goodin (1976); Sherman (1978)). These schemes all recognize that there are observation errors in the horizontal winds and attempt to adjust the wind field to generate mass conservation while maintaining the general sense of the large-scale flow field. Unfortunately, such procedures can result in vertical velocities that may be more representative of errors in the observed horizontal winds than of the actual vertical motion.

Some success has been achieved in developing procedures whereby the divergence at each grid point is reduced. On the regional and global scales, the simulation time period of interest is quite long (days to weeks to months), and the residual divergence in the flow must be correspondingly decreased to minimize inaccurcies from this source. For example, if one is content with a 10% error introduced by this process, residual divergences at each grid point must be less than 10^{-6} s^{-1} for a 1-day simulation, 10^{-7} s^{-1} for a 10-day simulation, and 10^{-8} s^{-1} for a 100-day simulation. This places great strains on the wind-adjustment scheme, especially when it should also not significantly distort the flow patterns. Current schemes probably reduce the residual divergence to a level appropriate for the 5 to 10-day simulation time.

TABLE 1
Summary of Inputs Needed to Solve the Atmospheric Diffusion Equation [Equation (1)].
(From Peters and Carmichael (1983))

Input	Level of Detail	Sources of Error
Wind velocities	Variation of **v** with **x** and t	Measurements available only at a few locations, generally at ground level, at discrete times; error in (1) variation of winds horizontally, (2) variation of winds vertically, and (3) determination of vertical velocity.
Eddy diffusivities	Variation of **K** with **x** and t	Few direct measurements available; **K** must be inferred from theory; both magnitude and vertical variation uncertain.
Chemical reaction mechanism	Rate equation for each $\langle c_i \rangle$	Inaccurate rates because of (1) inability to simulate atmosphere in laboratory, (2) unknown reactions, and (3) unknown rate constants
Source emissions	Emission rate as a function of **x** and t	Inaccurate knowledge of (1) variability of source activity and (2) emission factors.
Boundary conditions	Location of upper vertical boundary as a function of position and t	Lack of data or adequate model of temperature structure of atmosphere
	At surface, need mass-transfer coefficient and concentration of absorbing gas	Lack of data on the concentration in the absorbing phase and incomplete knowledge of deposition process

An alternate approach to eliminate these effects uses a divergence correction formulation. This simply subtracts the value of the divergence at each grid point from

Equation (1). Thus,

$$\frac{\partial \langle c_i \rangle}{\partial t} + \nabla \cdot (\langle \mathbf{v} \rangle \langle c_i \rangle) - (\nabla \cdot \mathbf{v}) \langle c_i \rangle = \nabla \cdot \mathbf{K} \cdot \nabla \langle c_i \rangle + \langle R_i \rangle + \langle S_i \rangle - \langle G_i \rangle .$$ (18)

The effectiveness of Equation (18) has been demonstrated by Kitada, Carmichael, and Peters (1984).

5 MODEL SIMULATIONS

In this section, results using an Eulerian combined transport/chemistry/removal model that describes the regional distribution of SO_2 and sulfate in the eastern United States are discussed. This model simulation incorporates the following significant features:

1. Gas-phase and heterogeneous chemistry varying diurnally, including reaction schemes to photochemically generate OH, HO_2, H_2O_2, and other radical concentrations;

2. Mixing layer development and dissipation, which enables emissions above the layer with possible subsequent entrainment later in the day and vice versa (i.e., emissions within the mixing layer may be transported aloft also);

3. Spatially and temporally varying wind, temperature, water vapor, and eddy diffusivity fields; and

4. Removal of SO_2 and sulfate at ground level with deposition velocity parameterized to depend on meteorological conditions, land-use type, and surface roughness.

Simulation results, including ambient concentration fields, deposition rates for SO_2 and sulfate, and a regional sulfur balance, were calculated for the period July 4-10, 1974. For this period, the synoptic situation over the eastern half of the United States was characterized by a high pressure region over the Atlantic Ocean. Surface winds were light and were generally from the southwesterly direction. High daily average ambient temperatures and dew point temperatures were observed during this period.

Predicted 24-hour average ground level SO_2 concentrations are presented in Figures 4 to 6 for July 7, 8, and 9. These contour plots show high ground level concentrations near the high source regions such as Chicago, St. Louis, western Pennsylvania, and the Ohio River Valley. Comparison of the ground level SO_2 concentration plots at 12:00 noon and 12:00 midnight (not shown here) indicated that the predicted nighttime concentrations were lower than the predicted daytime concentrations. This was probably due to the stable stratification prevalent at night, with the result that emissions from elevated sources were mixed to ground very slowly.

Predicted 24-hour average sulfate concentrations at ground level are shown in Figures 7 to 9 for July 7, 8, and 9. The effect of horizontal transport can be seen by observing the change in direction of the sulfate contours near western Pennsylvania. High sulfate concentrations predicted for July 7 are partly due to high ambient temperatures (> 24° C) and high average dew point temperatures during that period. For example, the predicted sulfate concentration at Wheeling, WV for July 6 (temperature < 24° C and dew point < 16° C) was 11.9 µg/m^3, whereas it was 17.2 µg/m^3 for July 7. The highest predicted SO_2 concentration during the period occurred near the Chicago area, and the highest sulfate concentration occurred around Wheeling, WV.

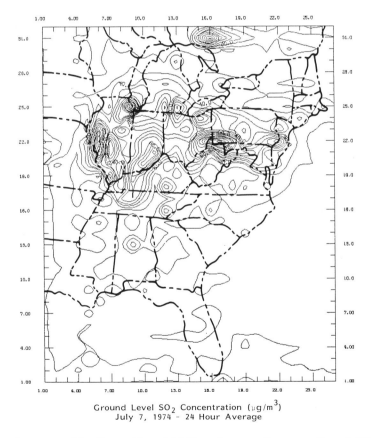

Ground Level SO_2 Concentration (µg/m^3)
July 7, 1974 - 24 Hour Average

FIGURE 4 Contours of the Ground Level 24-Hour Average Concentration of SO_2 for July 7, 1974.

The daily accumulated dry deposition contours for SO_2 and sulfate closely followed the ground level SO_2 concentration pattern as was reported by Carmichael and Peters (1984a,b). Thus, the SO_2 dry deposition contours were source-dominated as were the concentrations.

Regional daily mass balances showed that on average the amount of SO_2 deposited exceeded the sulfate deposited by eight times, and the amount of SO_2 deposited exceeded the amount reacted by nearly an order of magnitude. For sulfate, the amount deposited and the amount formed by reaction were of the same order of magnitude.

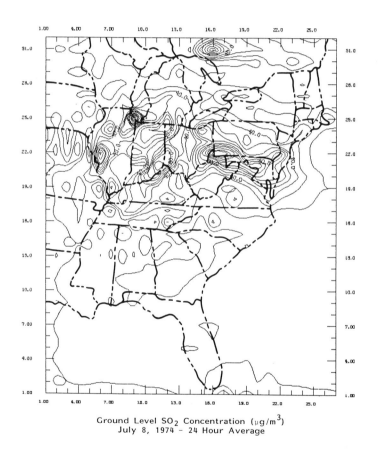

Ground Level SO_2 Concentration $(\mu g/m^3)$
July 8, 1974 – 24 Hour Average

FIGURE 5 Contours of the Ground Level 24-Hour Average Concentration of SO_2 for July 8, 1974.

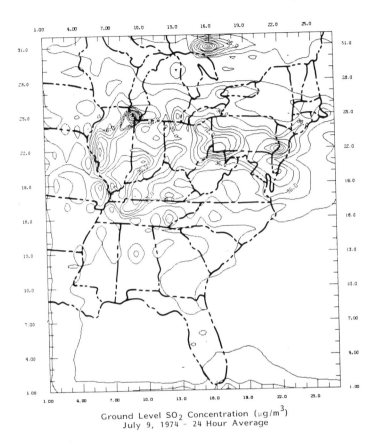

Ground Level SO$_2$ Concentration (μg/m^3)
July 9, 1974 – 24 Hour Average

FIGURE 6 Contours of the Ground Level 24-Hour Average Concentration of SO$_2$ for July 9, 1974.

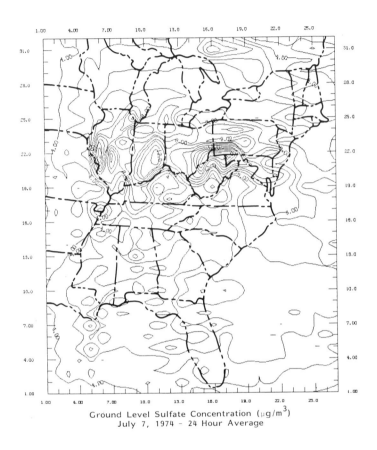

Ground Level Sulfate Concentration ($\mu g/m^3$)
July 7, 1974 - 24 Hour Average

FIGURE 7 Contours of the Ground Level 24-Hour AVerage Concentration of Sulfate for
July 7, 1974.

23

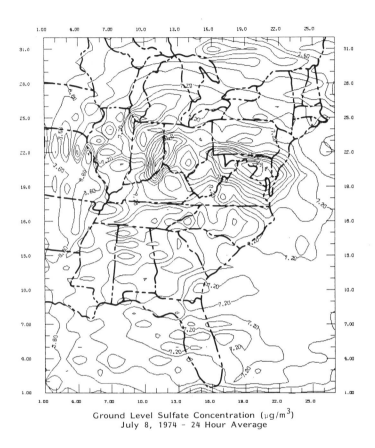

Ground Level Sulfate Concentration ($\mu g/m^3$)
July 8, 1974 - 24 Hour Average

FIGURE 8 Contours of the Ground Level 24-Hour Average Concentration of Sulfate for
July 8, 1974.

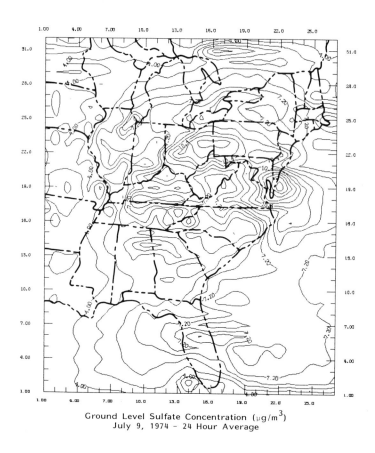

Ground Level Sulfate Concentration $(\mu g/m^3)$
July 9, 1974 - 24 Hour Average

FIGURE 9 Contours of the Ground Level 24-Hour Average Concentration of Sulfate for July 9, 1974.

6 CONCLUDING REMARKS

In this paper we have discussed several important aspects of atmospheric transport/chemistry model development. The presentation has not included all the significant phenomena occurring, but we have tried to emphasize some of the processes that are common to many species. Regional- and global-scale transport/chemistry models of trace species will be an important area of atmospheric sciences over the next decade, as we try to increase our knowledge and understanding of the impact of anthropogenic sources of pollution on the environment.

Our recent work has included development of a second-generation Eulerian model, primarily directed at analyzing regional-scale transport and deposition of O_3, SO_x, NO_x, and other trace gas species. The major advances relative to the initial regional-scale model are expanded and improved chemistry, incorporation of wet deposition and heterogeneous chemical processes, and improved numerics. Some of these aspects have been discussed in this review.

The model and its components have, thus far, been used for a variety of simulation applications. Some of these are (a) transport of SO_2/NO_x/hydrocarbon urban plumes to the background troposphere (Balko and Peters (1983)); (b) sulfate and nitrate formation in the presence of a land/sea breeze (Kitada, Carmichael, and Peters (1984)); (c) effects of in-cloud and below-cloud scavenging on homogeneous gas-phase chemistry (Carmichael, Kitada, and Peters (1983)); and (d) detailed analysis of mixing-limited chemical reaction of NO emissions (Karamchandani and Peters (1987)). The model is also being used to analyze field data, to evaluate emission control strategies, and to investigate local flow situations.

Current development work is emphasizing subgrid-scale parameterization of sources, chemical reactions, and cloud processes. Subgrid-scale processes are important in driving regional- and global-scale transport and chemical processes. This area has received relatively little attention, although one can readily identify a number of important problem areas. For example, since considerable chemical conversion may occur on a time scale typical of the transport over one grid length, the effective source strength that should be used in the Eulerian analysis can be significantly reduced. Second, large-scale plumes that originate from different emission sources can interact on the subgrid-scale and substantially alter the effective source distribution. Finally, many sources of anthropogenic pollution are emitted from so-called point sources. It has been conventional to sum these point source strengths over the grid volume and replace the sum by the volume-averaged value. Although this is convenient, it introduces errors into the analysis, and there appears to have been little consideration given to this source of error.

7 ACKNOWLEDGEMENTS

The research described in this paper has benefited greatly by discussions with my colleagues Dr. Gregory R. Carmichael and Dr. Toshihiro Kitada, as well as a number of graduate students. This work was supported in part by the National Aeronautics and Space Administration under research Grant NAG 1-36, and the National Acid Precipitation Assessment Program of the U.S. Environmental Protection Agency.

8 REFERENCES

Alkezweeny, A.J. and Powell, D.C., 1977. Estimation of transformation rate of SO_2 and SO_4 from atmospheric concentration data. Atmos. Environ., 11:179-182.

Balko, J.A. and Peters, L.K., 1983. A modeling study of SO_x - NO_x - hydrocarbon plumes and their transport to the background troposphere. Atmos. Environ., 17: 1965-1978.

Carmichael, G.R., Kitada, T., and Peters, L.K., 1983. The effect of in-cloud scavenging on the transport and gas phase reactions of SO_x, NO_x, HC_x, H_xO_y, and O_3 compounds. In Precipitation Scavenging, Dry Deposition, and Resuspension, H.R. Pruppacher, R.G. Semonim, and W.G.N. Slinn, Eds., Elsevier, Vol. 1, pp. 675-686.

Carmichael, G.R. and Peters, L.K., 1980. The transport, chemical transformation, and removal of SO_2 and sulfate in the eastern United States. In Atmospheric Pollution 1980, Studies in Environmental Science, Vol. 8, M.M. Benarie, Ed., Elsevier, Amsterdam, pp. 31-36.

Carmichael, G.R. and Peters, L.K., 1984a. An Eulerian transport/transformation/removal model for SO_2 and sulfate - I. Model development. Atmos. Environ., 18:937-951.

Carmichael, G.R. and Peters, L.K., 1984b. An Eulerian transport/transformation/removal model for SO_2 and sulfate - II. Model calculation of SO_x transport in the eastern United States. Atmos. Environ., 18: 953-967.

Donaldson, C.D. and Hilst, G.R., 1972. Effect of inhomogeneous mixing on atmospheric photochemical reactions. Environ. Sci. Technol., 6 :812-816.

Forester, C.K., 1977. Higher order monotonic convection difference schemes. J. Comput. Phys., 23 :1-22.

Goodin, W.R., McRae, G.J., and Seinfeld, J.H., 1980. An objective analysis technique for constructing three-dimensional urban-scale wind fields. J. Appl. Meteorol., 19: 98-108.

Hegg, D.A. and Hobbs, P.V., 1981. Cloud water chemistry and production of sulfates in clouds. Atmos. Environ., 15:1597-1604.

Heinrich, J.C., Huyakorn, P.S., Zienkiewicz, O.C., and Mitchell, A.R., 1977. An "upwind" finite element scheme for two dimensional convective transport equation. Int. J. Num. Meth. Eng., 11: 131-143.

Karamchandani, P.K. and Peters, L.K., 1987. Three-dimensional behavior of mixing-limited chemistry in the atmosphere.. Atmos. Environ., in press.

Kasibhatla, P.S., 1984. Some aspects of regional scale air pollution modeling. M.S. Thesis, Department of Chemical Engineering, Univeristy of Kentucky.

Kewley, D.J., 1978. Photochemical ozone formation in the Sydney airshed. Atmos. Environ., 12: 1895-1900.

Kitada, T., Carmichael, G.R., and Peters, L.K., 1983. The locally-one-dimensional, finite element method (LOD-FEM) for atmospheric transport/chemistry calculations. Proceedings of the Third International Conference on Numerical Methods in Engineering, Paris, France, March, 10 pages.

Kitada, T., Carmichael, G.R., and Peters, L.K., 1984. Numerical simulation of the transport of chemically reactive species under land- and sea-breeze circulations. J. Climate Appl. Meteor., 23: 1153-1172.

Lewellen, W.S. and Teske, M., 1975. Turbulence modeling and its application to atmospheric diffusion. Part I: Recent program development, verification, and application. U.S. EPA Report EPA-600/4-75-0169; National Technical Information Service PB-253450.

Liu, C.Y. and Goodin, W.R., 1976. An iterative algorithm for objective wind field analysis. Monthly Weather Rev., 104: 784-792.

Mahlman, J.D., Levy, H., and Moxim, W.J., 1980. Three-dimensional tracer structure and behavior as simulated in two ozone precursor experiments. J. Atmos. Sci., 37: 655-685.

McRae, G.J., Goodin, W.R., and Seinfeld, J.H., 1982. Numerical solution of the atmospheric diffusion equation for chemically reactive flows. J. Comput. Phys., 45: 1-42.

Mitchell, A.R., 1969. Computational Methods in Partial Differential Equations, John Wiley and Sons, New York.

Peters, L.K., 1983. Gases and their precipitation scavenging in the marine atmosphere. In Air-Sea Exchange of Gases and Particles, P.S. Liss and W.G.N. Slinn, Eds., Reidel Publishing Co., Dordrecht, pp. 173-240.

Peters, L.K. and Carmichael, G.R., 1983. Modeling of transport and chemical processes that affect regional and global distributions of trace species in the troposphere. In Trace Atmospheric Constituents: Properties, Transformations, and Fates, S.E. Schwartz, Ed., John Wiley and Sons, New York, pp. 493-538.

Peters, L.K. and Jouvanis, A.A., 1979. Numerical simulation of the transport and chemistry of CH_4 and CO in the troposphere. Atmos. Environ., 13: 1443-1462.

Preussner, P.R. and Brand, K.P., 1981. Application of a semi-implicit Euler method to mass action kinetics. Chem. Eng. Sci., 10: 1633-1641.

Seinfeld, J.H., 1975. Air Pollution: Physical and Chemical Fundamentals, McGraw-Hill, New York, p. 268.

Shannon, J.V., 1981. A model of regional long-term average sulfur atmospheric pollution, surface removal, and net horizontal flux. Atmos. Environ., 15, 689-702.

Sherman, C.A., 1978. A mass-consistent model for wind fields over complex terrain. J. Appl. Meteorol., 17: 312-319.

Stockwell, W.R. and Calvert, J.G., 1983. The mechanism of the HO - SO_2 reaction. Atmos. Environ., 17: 2231-2235.

Tarbell, T.C., Warner, T.T., and Anthes, R.A., 1981. An example of the initialization of the divergent wind component in a mesoscale numerical weather prediction model. Monthly Weather Rev., 109: 77-95.

ATTENTUATION OF AIR POLLUTION BY GREEN BELT

RAMESH K. KAPOOR and V.K. GUPTA
Safety Review Committee, Department of Atomic Energy, Bhabha Atomic Research Centre, Trombay, Bombay-400 085, INDIA

1 ABSTRACT

Plantation of green belt, around air pollution sources as well as around monuments of historical significance, has been suggested by many workers. However, in the absence of any mathematical models, these suggestions have mostly been qualitative in nature. A model for estimation of pollution attenuation for a green belt around a ground level pollution source has been developed by the authors. The model introduces the concept of pollution attentuation coefficient for estimating the removal of pollutant while passing through the green belt. The formulation for pollution attentuation coefficient makes use of widely measured parameters such as leaf area density of trees, deposition velocity of the pollutant on the leaf surface and wind speed in the green belt. The mathematical expression for estimation of the pollution attenuation factor of the green belt around a ground level population source uses a combination of an exponential law for dry removal of pollution within the green belt and source depletion model for removal outside the green belt. The formulation for population attenuation coefficient is tested for particulate material using the experimental data available in the literature. The dependence of the pollution attenuation factor on various physical parameters of the green belt such as its height, width, distance from the pollution source and on atmospheric stability conditions is illustrated. The model has been found to be useful to optimise the design of the green belt, to obtain the desired degree of attenuation of the pollution around an industry.

The model has been applied for two cases viz. reducing the consequences of reactor accidents by green belt and effectiveness of green belt to protect the Taj Mahal against air pollution. It is seen from these studies that the green belt is useful in reducing the consequences of a reactor accident involving releases at ground level, by orders of magnitude, whereas, its usefulness for protecting the Taj Mahal is found to be limited.

The development of the model and its application to the two cases mentioned above is presented. The paper also discusses the limitation of the model and suggests areas in which further research needs to be carried out.

2 INTRODUCTION

The pollution removal property of vegetation has been known for a long time. Trees and other vegetation interact with both the gaseous and particulate pollutants and remove them. For several years tree planting has been promoted by USSR scientists and city planners for the purpose of reducing ground level air pollution (Bennet and Hill (1975)). In the past, many workers (Kalyushnyi et al. (1952); Flemming (1967);

Bernatzky (1968); Hanson and Throne (1970); Warren (1973); Ganguly (1976)) have suggested the use of a green belt (rows of trees) for reducing the spread of pollution originating from industrial operations. These ideas lacked the quantitative treatment of pollution removal capacity of vegetation. Experimental work has been carried out by many workers, using field and laboratory experiments, to quantitatively assess the pollution uptake rates of vegetation. However, the findings of these experiments could not be used for the quantitative assessment of attenuation of air pollution by green belt until recently. This is mainly because, the results of these experiments on uptake of pollutants by plants have been expressed in terms of deposition velocity. The concept of deposition velocity, though suitable (for estimating the loss of pollutants over a vegetation surface (Hosker (1973)), is hard to apply for the case when pollutants travel horizontally through the plant canopies.

A theoretical model recently developed by Kapoor and Gupta (1984) estimates the pollution attenuation factor of a green belt around a ground level pollution source. The model uses a combination of an exponential law for dry removal of pollution within a plant canopy and a source depletion model outside the canopy. A model for green belt for elevated releases is not yet available in the literature. Since ground level pollution releases under normal and accidental conditions are more harmful than elevated releases, a suitably designed green belt around such pollution sources can provide an added buffer for public safety.

As a number of review papers on interaction of pollution with vegetation are available in literature (Hosker and Lindberg (1982); Shinn (1979)); and books edited by Mudd and Kozlowski (1975); and by Monteith (1975) give a number of review papers on this topic), these are not reviewed in this paper; instead a review of the work of authors of this paper carried out on the use of green belt for attenuation of pollutants is given here. This paper reviews the theoretical model for green belt and its application for a few cases. The applications include both the source and receptor oriented approaches. Some ideas on the scope of future work in this field are also presented.

3 MODEL
3.1. Exponential Removal Law

A series of experiments carried by Raynor et al (1974) at Brookhaven National Laboratory, USA using pollens and spores as tracers for studying the removal of paticulates by forest indicated that mass flux of tracer decreases nearly exponentially with travel distance within the forest. While introducing the concept of a filteration model to describe the fraction of depositing aerosol which is intercepted by the herbage when the released is from above, Chamberlain (1970) and Chadwick and Chamberlain (1970) proposed the exponential removal law for the removal of pollutants. The exponential law for dry removal of pollutants has also been suggested by USEPA (1978) to account for plume depletion in air quality models. Considering the exponential removal law, if the mass flux of a pollutant which enters the green belt is Qc, the mass flux Qx at a travel distance x within the green belt is given by

$$Qx = Qc \ exp(-\lambda x) \tag{1}$$

where $\lambda(m^{-1})$ is the pollution attenuation coefficient.

Experimentally measured values of λ are not yet available in the literature.

The following relationship for λ in terms of easily measurable parameters was proposed by Kapoor and Gupta (1984)

$$\lambda = k\rho_t \ V_d/Uc \tag{2}$$

where

V_d = dry deposition velocity of pollutant for vegetative canopy (ms^{-1});
U_c = average windspeed through the green belt (ms^{-1});
ρ_t = foliage surface area density of a single tree (m^2m^{-3}) ; and
$k = \rho_c/\rho_t$

where,

ρ_c = average foliage surface area density of the green belt (m^2m^{-3}) .

The constant k depends upon the spacing of trees in the green belt and is introduced to consider the overall foliage surface area density. For an ideal green belt having the same foliage surface area density as that of a single tree, the value of k will be unity. This formulation of λ is similar to equation (2) of Bache (1981) which was suggested as a canopy absorption coefficient.

3.2 Testing the Formulation of λ

Since λ can be estimated both from equations (1) and (2), the formulation for λ (Equation 2) can be tested provided, for the same set of field experiments, the values of λ as obtained from both of these equations agree well with each other. In the literature, such a complete experiment, where all the parameters of equations (1) and (2) are measured, is not available. One difficulty is that the measured values of V_d for plant canopies are not available in the literature. Hence, one is left with no other choice than to use either the value of V_d representative of grass, crops or a theoretically estimated value based on models developed for plant canopies (Bache (1979, 1981); Slinn, (1982)). Few field experiments have been done to which equation (1) can be applied. The only experiments of which we are aware that can be used for this purpose are those carried by Raynor et al (1974) using spores and pollens. In view of these difficulties, the formulation can at best be tested for these experiments.

Using the experimental data of Raynor et al., the average value of λ for Ragweed pollen (size 20 μm) as obtained from (1) is 0.0187 m^{-1} ± 0.0047 m^{-1}. This value can be compared with the value of λ obtained from (2) provided the appropriate values of k, ρ_t, V_d, and U_c applicable to the experimental site of Raynor et al are used. The details of obtaining these values have been discussed by Kapoor and Gupta (1984). Using these values in (2), the value of λ for the experimental site of Raynor et al. ranges from 0.0088 to 0.0181 m^{-1} which compares well with the value of λ as obtained from (1). Considering the uncertainties associated with k, ρ_t, U_c and V_d, the formulation of λ given in (2) seems to be a reasonably valid approximation for depletion of Ragweed pollen through a pine forest. Additional studies need to be carried out for different pollutants and trees to test the validity of λ, since the authors have tested it only for the case Ragweed/pine tree data.

3.3. Attenuation Factor of Green Belt

A schematic of the green belt around an industry is given in Figure 1. The effectiveness of a green belt in attenuating the pollution is given by the value of attenuation factor A_f which is defined as the ratio of mass flux of pollutant reaching at distance $(x_1 + x_2)$ in the absence of the green belt, Q_{WB}, to the mass flux reaching at the same distance in the presence of the green belt, Q_B, and is given by

$$A_f = Q_{WB}/Q_B \tag{3}$$

The value of A_f is calculated in the following five stages.

Stage 1: Mass flux Q_A reaching at the inner edge of green belt. The depletion due to dry deposition over the travel distance up to the inner edge (x_1) of the green belt and above it (x_2) is obtained using the source depletion model (Chamberlain (1953)). The mass flux Q_A reaching the location A can be evaluated by

$$Q_A = Q_0\, F_D(x_1) \tag{4}$$

where Q_0 is the mass flux at the source, and

$$F_D = [\exp \int_0^x \frac{1}{\sigma_z} \exp(-\frac{H^2}{2\sigma_z^2})\, dx]^{-(2/\pi)^{1/2}} (V_d/U) \tag{5}$$

where H is the release height, σ_z is the standard deviation of concentration distribution in vertical direction, V_d is the dry deposition velocity, U is the mean wind speed and x is the travel distance.

Value of F_D can be obtained either by the above integration or from tables given in the literature (e.g. Shirvaikar and Abrol (1978) which give the value of F_{D0} for ratio of $V_d/U = 0.01$ at various distances under different atmospheric stability conditions). The value of F_D for any ratio of V_d/U is given by

$$F_D = (F_D)^{100 \; V^d /U} \qquad (6)$$

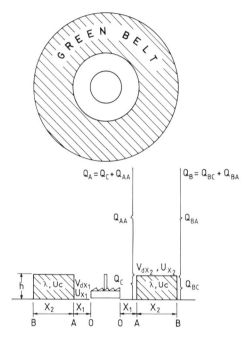

FIGURE 1 Diagram of a green belt around a ground level pollution source illustrating the parameters of pollution attenuation model.

The value of Q_A can thus be obtained from (4) and (6) .

Stage 2: Fraction of mass flux Q_A entering the green belt. The material Q_A is divided into two parts, namely Q_c which passes through the green belt and Q_{AA} which passes over it. The proportion in which Q_A is divided into these two parts depends upon the height of trees, distance x_1 and atmospheric conditions. The effective height (h_e) of the incident air stream which enters the green belt of height h is given by

$$\int_0^{h_e} U(z) \, dz = h \, U_c \qquad (7)$$

where $U(z)$ is the wind speed profile outside the green belt. The methodology for calculation of h_e is given by Kapoor and Gupta (1984). The quantity Q_c is then evaluated as follows:

$$Q_c = \int_0^{h_e} \int_{-\infty}^{\infty} CU \, dy \, dz \qquad (8)$$

where concentration C is given by

$$C = \frac{Q_0}{\pi \, \sigma_y \, \sigma_z \, U} \exp(-\frac{z^2}{2\sigma_z^2}) \exp(-\frac{y^2}{2\sigma_y^2}) \qquad (9)$$

where σ_y and σ_z are the standard deviations of concentration distribution in cross wind and vertical directions respectively at down wind distance x_1. Equation (8), after carrying out the integration over y and z, can be written as

$$Q_c = Q_A \, \text{erf}(\frac{h_e}{\sqrt{2} \, \sigma_z}) \qquad (10)$$

where,

$$\text{erf}(x) = \frac{2}{\sqrt{\pi}} \int_0^x e^{-t^2} dt \qquad (11)$$

The value of Q_{AA} can be obtained by the following relation

$$Q_{AA} = Q_A - Q_c \qquad (12)$$

Stage 3: Mass flux Q_{BC} coming out of green belt. The material Q_c is depleted in the green belt according to the following relationship

$$Q_{BC} = Q_c \exp(-\lambda \, x_2) \qquad (13)$$

where x_2 is the width of the green belt. The amount of material Q_{AA} which passes over the green belt will also get depleted to mass flux Q_{BA} by loss of material at the top surface of the green belt. The value of Q_{BA} is evaluated for travel distance x_2 from (4) and (6) by replacing Q_0 by Q_{AA}.

Stage 4: Total mass flux Q_B reaching at the outer edge of green belt. The total mass flux Q_B reaching at the outer edge (location B) of the green belt is thus given by

$$Q_B = Q_{BC} + Q_{BA} \qquad (14)$$

Stage 5: Calculation of Attenuation Factor A_f . In the absence of a green belt, the mass flux of Q_{WB} reaching the distance $x_1 + x_2$ is evaluated using equations (4) and (6). The attenuation factor A_f of the green belt at B is obtained using (3).

Combining all the five above stages, the pollution attenuation factor A_f is finally given by

$$A_f = F_D(x_1 + x_2)/F_D(x_1)[\text{erf}(x)\, e^{-\lambda x_2} + \text{erfc}(x)\, F_D(x_2)] \tag{15}$$

where

$F_D(x_1 + x_2)$ = plume depletion factor for travel over distance $x_1 + x_2$,
$F_D(x_1)$ = plume depletion factor for travel over distance x_1
$F_D(x_2)$ = plume depletion factor for travel over distance x_2 .

An illustrative example of optimization of green belt using this model for a few selected values of parameters is given in the following section.

3.4. Optimization of Green Belt

The case of an industry surrounded by a green belt as given in Figure 1 is used here for estimating the optimum size of green belt. The release of pollutants near the ground level is assumed.

The estimated value of attenuation factor A_f obtained by using equation (15) for combinations of green belt parameters x_1, x_2, h, λ, U under different atmospheric stability conditions can be used to optimize the design of the green belt. The parameters x_1, x_2 and λ are critical for optimization. The choice of the remaining parameters (h and U) is restricted and is decided by the local conditions. For example, it is preferable to have taller trees with high foliage density, but trees of height greater than 20 m are not common at most sites. As an example, the calculations are performed for removal of particulate material by a green belt for $\lambda = 0.02$ m^{-1} and tree height varying from 10 to 30 m. Among the other parameters, values of V_{dx1} and V_{dx2} are assumed to be equal and assigned a value of 0.0156 s^{-1} (which is the gravitational settling rate for 20 μm particle size). Similarly, U_{x1} and U_{x2} are assumed to be the same and assigned values as given below under different atmospheric stability conditions:

Stability Condition:	A	B	C	D	E	F
Wind speed (ms^{-1}):	2	2	4	5.5	3	2

Variation of A_f with x_1 and x_2

For a fixed height of trees, the value of A_f is expected to increase with either an increase in x_2 or with a decrease in x_1. Though it is preferable to plant the green belt as close as possible to the pollution source (i.e. low value of x_1), practical considerations may not permit very low values of x_1. Similarly, for fixed values of x_1 and x_2, the value of A_f is larger for taller trees. The variation of A_f with x_1, x_2 and h under different atmospheric stability conditions is discussed below.

Unstable Condition A: The variation of A_f with x_1 for a fixed value of $x_2 = 500$ m and with x_2 for a fixed value of $x_1 = 50$ m for stability condition A are given in Figures 2 and 3 respectively. It can be seen from Figure 2 that the value of A_f is close to unity even for taller trees if x_1 is 300 m or more. Reducing the value of x_1 is of advantage only in the case of tall trees and a value of $A_f = 6$ is obtained for 30 m tall trees if x_1 is kept 50 m.

Figure 3 shows that the value of A_f remains constant beyond a certain value of x_2 depending on the height of trees. This suggests that increasing the value of x_2 more than 300 m does not serve any useful purpose for stability condition A.

Stable Condition F: Figures 4 and 5 are similar to Figures 2 and 3 but present the variation of A_f in stability condition F. The value of A_f, as high as 10^4, is attainable in this stability condition (see Figure 4) for trees of height 15 m or more for $x_2 = 500$m and $x_1 = 50$m. This figure also shows that if the tree height is more than 25m, the value of A_f is of the same order even if x_1 is increased to 100m. Figure 5 shows that the value of A_f increases with x_2 up to a certain value of x_2 and thereafter the increase in A_f is insignificant. This feature, relatively constant A_f beyond a certain x_2, helps in estimating the optimum width of the green belt. For the conditions considered in this example, this width is 500m, 700m, and 1700m for tree heights 10m, 15m, and 30m respectively.

Variation in Other Atmospheric Stability Conditions

The above discussion pertains to only two extreme atmospheric stability conditions. The values of A_f for intermediate stability conditions lie mostly between the two extreme values (Table 1). It can be seen from this table that for 15m high trees, advantage gained by increasing the width of green belt from 500 to 1000m is not significant in all the stability conditions except for F where A_f of more than 10^4 is obtained for $x_1 = 50$m. However, if the tree height is increased to 30m, an attenuation factor of greater than 10^4 is possible in stability category E also for a green belt width of 1000m, whereas, in stability category D , the attenuation factor obtained is 600 . This suggests that a green belt consisting of tall trees around an industry could provide significant attenuation factors for about 50% of the time in a year for the case considered here.

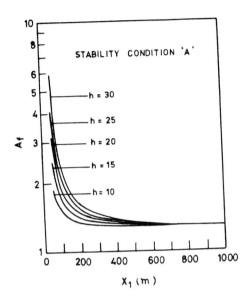

FIGURE 2 Attenuation factor as a function of distance between the source and inner edge of green belt and the tree height, for $x_2 = 500$m, $\lambda = 0.02$ m^{-1} and for atmospheric stability condition A .

TABLE 1

Computed Values of A_f for Different Atmospheric Stability Conditions for $x_1 = 50$ m and $\lambda = 0.02$ m^{-1}

x_2 (m)	h (m)	A	B	C	D	E	F
500	15	2.5	3.6	4.5	1.7×10^1	1.2×10^2	1.1×10^4
	30	6.2	1.4×10^1	2.6×10^1	6.0×10^2	1.4×10^4	1.1×10^4
1000	15	2.5	3.6	4.5	1.7×10^1	1.2×10^2	5.8×10^5
	30	6.2	1.5×10^1	2.6×10^1	6.0×10^2	1.1×10^5	8.2×10^7

Values of "A_f" for Stability Classes

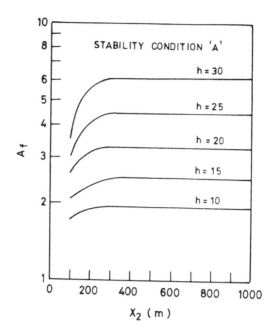

FIGURE 3   Attenuation factor as a function of width of green belt and the tree height, for $x_1 = 50m$, $\lambda = 0.02m^{-1}$ and for atmospheric stability condition A .

4 APPLICATION EXAMPLES OF GREEN BELT MODEL

Two examples of application of the green belt model are given here to illustrate the usefulness of the green belt in reducing the impact of pollution. The first example concerns the reduction of the consequences of nuclear power reactor accidents and the second is on protection of the Taj Mahal against air pollution by green belt. The former is source oriented while the latter is a receptor oriented approach. These examples suggest that a green belt is more effective in attenuating air pollution where it is used in a source oriented manner. Both of these approaches are discussed in the following sections.

4.1. A Source Oriented Example

A nuclear power reactor surrounded by a suitably designed green belt (Figure 6) provides an example of the source oriented approach.

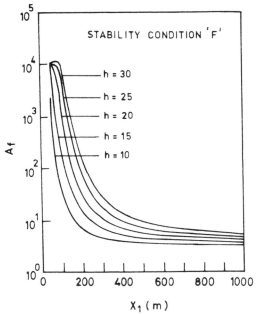

FIGURE 4   As for Figure 2 but for atmospheric stability condition F .

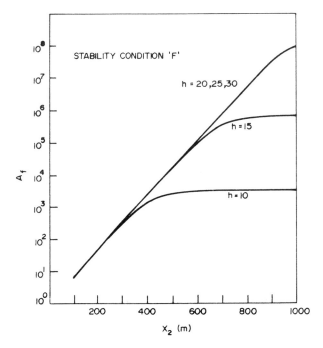

FIGURE 5 As for Figure 3 but for atmospheric stability condition F .

40

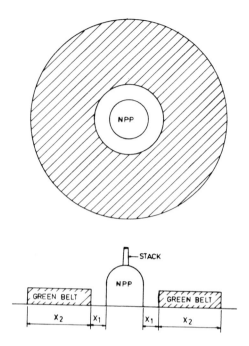

FIGURE 6 Schematic of a green belt of trees around a Nuclear Power Plant (NPP)

In the nuclear power industry, low probability - high consequence events have attracted special attention. Safety considerations are mainly directed towards reducing the probability of these events by adopting the "Defense-in-Depth" policy in design, construction, and operation of the reactor. These are coupled with viable "Emergency Preparedness Plans" to reduce the consequences of accidents. Presence of a green belt around a nuclear power reactor can reduce the consequences of reactor accidents by absorbing significant amounts of the radioactive materials before they reach the public. Gupta and Kapoor (1985) analysed the consequences of a hypothetical reactor accident, in which a large, cold, ground level release of radioactive nuclides takes place in the presence of a suitably designed green belt around a nuclear power plant. Release of 54 radionuclides from a 1000 MWe light water reactor was considered. These radionuclides fall into two main categories: particulates and noble gases. The deposition velocity on vegetation for particulates was taken as 0.01 ms^{-1} and zero for noble gases (being inert in nature). The calculations considered the building wake effect and radioactive decay as well as deposition of radionuclides in the green belt.

The radiological consequences were developed in terms of early and continued mortalities. The radiation dose to four important body organs (viz. thyroid, lung, gastrointestinal tract, and bone marrow) of persons staying up to 100 kms from the reactor were calculated under different atmospheric stability categories. The dose-mortality criteria of WASH 1400 (1975) were adopted for calculating early and continued mortality. The reactor accident consequences were calculated without a green belt and compared with the corresponding consequences when the reactor is surrounded by a 15m high green belt of 1500m width. The inner boundary of the green belt was taken as 50m away from the reactor.

The results are discussed for two extreme atmospheric stability categories, A and F. Of these two, the consequences of a ground level release are maximum in category F. Further, the consequences of exposure to bone marrow alone are discussed because it is the most sensitive of the organs.

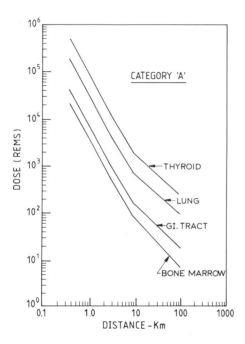

FIGURE 7 Total organ doses versus distance from a Nuclear Power Plant for atmospheric stability category A (without a green belt)

A comparison of dose to different organs from the radioactive release between the case of a reactor without a green belt and the case with a green belt is given in Figures 7 and 8 for the atmospheric stability category A. A similar comparison for category F is given in Figures 9 and 10.

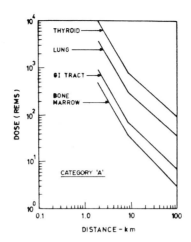

FIGURE 8 As for Figure 7 but with a green belt of 1500m width, height 15m located 50m away from NNP

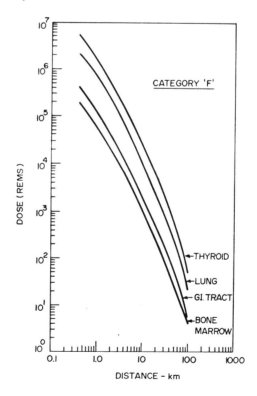

FIGURE 9 As in Figure 7 but for atmospheric stability category F

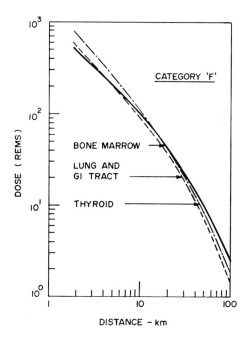

FIGURE 10 As in Figure 8 but for atmospheric stability category F .

      In category F , the 1500m wide green belt reduces the dose to bone marrow at 2 km distance by a factor of 40 ; whereas, in category A , this reduction factor is only 2 This dose reduction factor is much less than the value of pollution attenuation factor A_f (Table 2). This is as expected since the green belt attenuates only the particulate radionuclides and not the noble gases. This differential attenuation cf particulate radionuclides is further illustrated in Table 3. For category F , it can be seen from this table that most of the particulate radionuclides are absorbed by the green belt and hence greater than 98% of contribution to dose is from noble gases. Whereas, in category A , the contribution from noble gas is about 8% since only a small fraction of particulate matter is absorbed by the green belt due to large vertical spread of plume.

      One of the important parameters of severe radiological consequences of a reactor accident is the distance for $LD_{50/60}$ (= 510 rems lethal to 50% of the exposed population in 60 days). The distances for $LD_{50/60}$ with and without a green belt of different widths for all the six atmospheric stability categories are given in Table 4. This table shows that a population receives an $LD_{50/60}$ dose up to a distance of 14.6 km from the plant without a green belt. This distance is less than 2 km for a green belt of width 1500 m in atmospheric stability category F . Since the radiation exposure of bone marrow is the dominant mortality risk, the impact of a green belt in reducing the early and continued mortality (based on bone-marrow dose-mortality factor) is presented in Figure 11. Without a green belt the mortality probability in category F is unity up to a distance of about 12 km ; however, with a 1500m wide green belt, the mortality probability is zero at 3 km .

TABLE 2

Pollution Attenuation Factor A_f of Green Belt of Different Widths

Stability Category	A_f for Green Belt Width (m)		
	700	1000	1500
A	2.31	2.37	2.38
B	3.12	3.37	3.39
C	3.40	3.97	4.26
D	4.71	7.75	12.61
E	16.71	44.80	96.68
F	27.69	128.04	1792.59

TABLE 3

Total Dose and % Contribution from Noble Gases to Bone Marrow Without and With green Belt for atmospheric Stability Category A and F .

	Stability Category A				
	Without Green Belt			With Green Belt	
Distance (Km)	Total dose (rems)	% Contribution from Noble Gases	Total Dose (rems)	% Contribution from Noble Gases
2.0	1.12×10^3	3.75	4.93×10^2	8.49
5.2	2.06×10^2	4.00	9.13×10^1	9.04
7.6	1.09×10^2	3.91	4.82×10^1	8.82
12.4	5.07×10^1	3.60	2.24×10^1	8.14
18.0	2.88×10^1	3.29	1.27×10^1	7.50
30.0	1.67×10^1	2.72	7.26	6.25

(Table 3 continued)

Stability Category F

| | Without Green Belt | | With Green Belt | |
Distance (Km)	Total dose (rems)	% Contribution from Noble Gases	Total Dose (rems)	% Contribution from Noble Gases
2.0	2.14×10^4	2.36	5.16×10^2	97.74
5.2	3.84×10^3	5.49	2.13×10^2	99.05
7.6	1.84×10^3	7.81	1.45×10^2	99.39
12.4	7.10×10^2	12.03	8.58×10^1	99.59
18.0	3.27×10^2	16.15	5.30×10^1	99.71
30.0	1.04×10^2	23.80	2.49×10^1	99.82

TABLE 4

*$LD_{50/60}$ Distance for Exposure to Bone Marrow

Stability Category	$LD_{50/60}$ Distance (km)			
	Without Green Belt	With Green Belt of Width (m)		
		700	1000	1500
A	3.1	2.00	1.75	<2.0
B	5.0	2.75	2.55	2.45
C	5.9	2.9	2.60	2.55
D	7.9	3.05	2.3	<2.0
E	14.2	3.75	1.9	<2.0
F	14.6	3.9	2.5	<2.0

*$LD_{50/60}$ = 510 rems

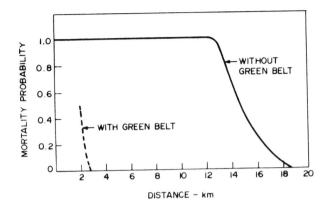

FIGURE 11 Mortality probability versus distance from NPP for bone marrow exposure for atmospheric stability category F (green belt width, 1500m).

Besides the decrease in mortality probability, the impact of green belt on reduction of consequences of lesser importance such as relocation of population and banning of food produced in contaminated regions were also calculated. The analysis shows that the benefits of developing a suitably designed green belt around nuclear power plants are as follows:

(i) The early and continued mortalities are practically eliminated beyond a distance of 3 km.

(ii) The relocation of population to protect it from long-term external exposure from ground contamination may not be required.

(iii) The supply of food from uncontaminated areas may be needed for populations living between 2 to 20 km only in the affected sector.

Since the siting requirements in many countries require exclusion distances ranging from 0.5 to 3 km from the reactor, it should be possible to devleop a green belt around nuclear power plants that will substantially reduce the consequences of the accidents. The green belt would thus serve as another barrier between the nuclear power plant and the public as well as strengthen the defense-in-depth philosophy adopted in the design, construction, and operation of nuclear power plants. The assured availability of a suitably designed green belt throughout the year could also help in reducing the magnitude of emergency preparedness in the public domain.

4.2 Receptor Oriented Example

The case of protecting the Taj Mahal against air pollution by a green belt is presented here as an example of a receptor oriented approach. Mathematically both approaches are the same. It is only in manner of the planting of green belt that the approaches differ. In the earlier source oriented approach, the green belt is planted around the pollution source, whereas in the receptor oriented approach the receptor is protected against air pollution by planting a green belt around the receptor.

Since the green belt model is developed for ground level sources only, its applicability to the case of the Taj Mahal which is affected also by elevated releases is questionable. It is pertinent to note here that the model, when applied to elevated sources, will overestimate the value of pollution attenuation factor A_f as comparatively less amount of polluted air will pass through the green belt. The extent of overestimation cannot be determined unless an appropriate model of green belt for elevated sources is developed. However, qualitatively, the overestimation of A_f is reduced for distant elevated sources under unstable atmospheric conditions. Thus it can be concluded that the value of A_f in reality for such a receptor oriented approach would be less than the values given here because the receptor is affected by elevated sources also.

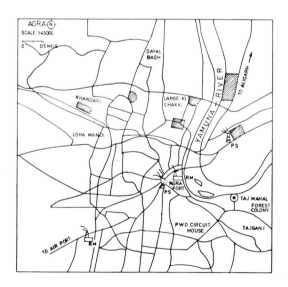

LEGEND:

INDUSTRY
PS - POWER STATION
RM - RAILWAY MARSHALLING YARD

FIGURE 12 Map of Agra showing industrial activity around the Taj Mahal

The major source of air pollution at Agra (Figure 12) affecting the Taj Mahal are as follows:

i) two thermal power plants consuming about 1100 tonnes of coal daily;

ii) industries including about 250 foundaries; the foundaries alone consume about 200 to 300 tonnes of coal daily;

iii) railway shunting yard close to Agra Fort consuming approximately 40 to 50 tonnes of coal daily;

and

iv) Besides these three major sources, the pollution is also generated by the use of coal, wood, dung cakes for domestic fuel which accounts for about 275 tonnes per day of coal equilvalent.

The value of A_f is calculated using the following equation:

$$A_f = F_D(x_1 + x_2)/F_D(x_1) \, [\mathrm{erf}(x) \, e^{-\lambda x_2} + \mathrm{erfc}(x) \, F'_D(x_2)] \tag{16}$$

Equation (16) is nearly the same as equation (15) except that here $F_D(x_2)$ has been replaced by $F'_D(x_2)$. This difference is mainly because, in the case of the Taj Mahal some of the pollution sources are far away from the receptor and thus resulting in a large vertical spread of that portion of plume which passes above the green belt. The value of $F'_D(x_2)$ is calculated assuming that the size of vertical spread above the green belt is obtained as if it has travelled over an imaginary distance x_i before arriving at the green belt. The size of the plume at the edge of the green belt is obtained from the following relationship:

$$\sigma_z(x_i) = \sigma_z(x_1) - h_e \tag{17}$$

and $F'_D(x_2)$ is obtained as follows:

$$F'_D(x_2) = F_D(x_i + x_2)/F_D(x_i) \tag{18}$$

where: $F_D(x_i + x_2)$ and $F_D(x_i)$ are the plume depletion factors due to dry deposition over $x_i + x_2$ and x_i respectively.

A schematic of the green belt around the Taj Mahal is given in Figure 13. Using the model described in the earlier section, a parametric study has been carried out to evaluate the effectiveness of the green belt around the Taj Mahal. The value of A_f is computed for different values of parameters as given in Table 5. The values of U, U_c

and h_e for different atmospheric stability categories are given in Table 6. The pollution attenuation factor A_f is calculated using these values of parameters.

TABLE 5
Values of parameters for computation of A_f

Parameter	Symbol	Values	Remarks
Ground Roughness (cm)	z_0	100	Representative of Urban area.
Height of Mixing Layer (km)	H_m	3.240	Annual average value for Delhi based on analysis carried out by Kapoor (1979)
Separation distance between source and Green Belt (Km)	x_1	0.5,1,2, 5,25,50	To cover a wide region of pollution sources.
Width of Green Belt (m)	x_2	100,200 500	Typical
Height of Green Belt (m)	h	10,20,30	Normal range of height of trees
Pollution attenuation coefficient (m^{-1})	λ	0.005,0.01, 0.02	Typical
Dry deposition velocity (ms^{-1})	V_d	0.01	Sehmel (1980)*

*deposition velocity of $1cms^{-1}$ is often assumed for reactive gases (such as SO_2 and Cl_2). However, it ranges two orders of magnitude around this value.

TABLE 6

Values of Parameters

Parameter	Stability Categories					
	A	B	C	D	E	F
$U(ms^{-1})$	2.0	2.0	4.0	5.5	3.0	2.0
$U_c(ms^{-1})$	0.5	0.5	0.8	1.0	0.6	0.5
he for h (m) (m)						
10	3.62	3.68	3.04	3.29	3.49	4.52
20	6.40	6.49	5.52	5.49	5.95	7.56
30	8.99	9.08	7.29	7.48	8.14	10.16

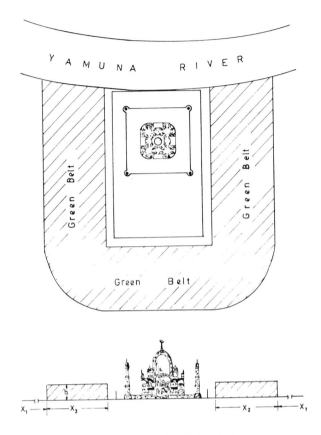

FIGURE 13 Schematic of green belt around the Taj Mahal

The variation of A_f with atmospheric stability categories for a typical set of values of parameters is given in Table 7. It can be seen that the value of A_f increases with increasing atmospheric stability. For atmospheric stability category F , for which the value of A_f is the highest among all the stability categories, the variation of A_f with x_1 for three different tree heights, $x_2 = 500m$ and $\lambda = 0.02 \ m^{-1}$ is given in Figure 14a. It is seen that A_f decreases with increasing value of x_1 and taller trees give higher attenuation factor for lower values of x_1 . For example, for $x_1 = 500$ m , the value of A_f increases from 1.7 to 5.2 if tree height is increased from 10m to 30m , whereas, for $x_1 = 50$ km , the corresponding value of λ viz 0.02 m^{-1} , the effect of changing the value of A_f increases from 1.018 to only 1.075. As Figure 14a corresponds to only one value of λ for category F , shown in Figure 14b, suggests that even for $x_1 = 500$ m the increase in value of A_f with increase in λ is not very significant. For higher values of x_1 , A_f is practically the same for the three values of λ .

TABLE 7

Computed Values of A_f Under Different Atmospheric Stability Categories

Atmospheric Stability Category	Value of A_f
A	1.0072
B	1.0143
C	1.0200
D	1.0300
E	1.0580
F	1.1871

Values of green belt parameters used in above computation of A_f are:

$x_1 = 5000$ m ; $x_2 = 500$ m ; $h = 20$ m ; $\lambda = 0.02$ m^{-1}

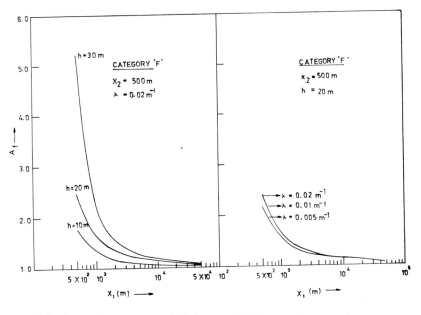

(a) Dependence on Tree Height   (b) Dependence on λ

FIGURE 14 Variation of A_f with X_1 in stability category F .

For atmospheric stability category A , for which the value of A_f is the least among all the categories, the variation of A_f with x_1 for three different heights of trees is given in Figure 15a. The shape of variation in this figure is similar to Figure 14a except that even for 30 m tall trees and $x_1 = 500$ m the value of A_f is only 1.07 in Category A   for $x_2 = 500$ m, in comparison to the corresponding value of A_f of 5.2 in Category F . The variation of A_f with x_1 for three values of λ for category A is shown in Figure 15b, which like category F shows very little dependence of A_f on value λ .

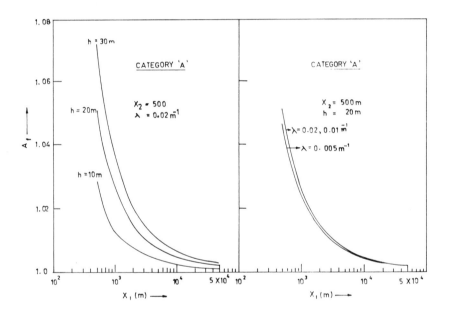

(a) Dependence on Tree Height   (b) Dependence on λ

FIGURE 15 Variation of A_f with X_1 in stability category A

The variation of A_f with the width of green belt (x_2) and its dependence on tree height and on λ for $X_1 = 500$ m is shown in Figures 16a and 16b respectively for the two extreme atmospheric stability categories viz. A and F . It can be seen that value of A_f increases with increases in x_2 only up to a certain extent and then remains practically independent of x_2 . This suggests that there is an optimum value of width of green belt beyond which A_f practically ceases to increase. An interesting feature of Figure 16b is that under atmospheric stability category F , unlike the variation of A_f with x_1 , the variation of A_f with x_2 is significantly dependent upon value of λ

particularly for smaller values of x_2. This dependence of A_f on λ in atmospheric stability category A is not very significant even for smaller values of x_2.

The results of this parameteric study of green belt suggest that the green belt is more effective in attenuating pollution in stable atmospheric stability conditions and for smaller distances between source and green belt (x_1). Increasing the height of trees is of advantage in attenuating the pollution only for smaller values of x_1. For the range of x_1 considered in this study, it is observed that by increasing the width of green belt more than 200 m does not significantly increase the value of A_f. For smaller widths of green belts, increasing the value of λ results in an increase in the value of A_f under stronger atmospheric stability. However, for a fixed value of x_1, x_2 and h, an increase in the value of λ from 0.005 to 0.02 m^{-1} does not significantly increase the value of A_f under unstable atmospheric conditions.

The results of this study indicate that in general, the value of A_f in the case of the Taj Mahal is much smaller than that obtained by a source oriented approach. Significant values of A_f occur only for the nearby ground level pollution sources and that also in stable atmospheric conditions. In view of the limited effectiveness of green belt in protecting the Taj Mahal, induction of pollution control technology at the source for industrial emissions and progressive replacement of polluting domestic fossil fuel by clean fuel seems to be an effective approach besides planting of a green belt around the Taj Mahal. Considering the topography around the Taj Mahal, it may not be possible to develop a green belt on the river side. It is suggested that this aspect be considered while developing the land area across the river and more emphasis be placed on the pollution control at the source level.

5 LIMITATIONS OF THE MODEL AND NEED FOR MORE RESEARCH WORK

The green belt model used here is applicable only for ground level releases of air pollution, and hence there is a need to develop a model for elevated releases. However, the green belt will be less effective for elevated releases since only a small fraction of pollutant will pass through the green belt. The green belt model has been tested only for ragweed/pine tree data. Additional field experiments need to be carried out to test the model for other pollutants and trees. For gaseous pollutants, since stomata on the leaves are the important means of uptake of pollutants by plants, the value of deposition velocity V_d would be a function of the numerous and varied factors which control the opening of stomata. Another unknown quantity is k in equation (2) which depends upon the spacing of trees. Studies need to be carried out for establishing the relationship of k in terms of spacing of trees. A keen look at equation (2) for λ indicates the value of λ can be increased by increasing the average foliage

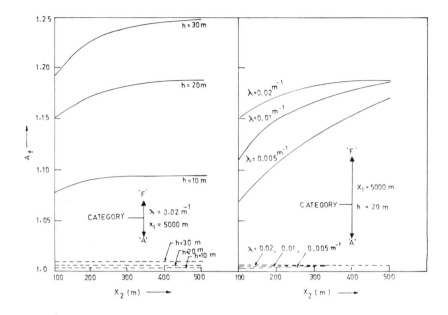

FIGURE 16 Variation of A_f with X_2 in stability categories A and F
(a) Dependence on Tree Height   (b) Dependence on λ

surface area density of the green belt. But this increase in foliage density would then reduce the wind speed through the green belt which results in decreased amount of pollutants entering the green belt. These two counter factors would then result in an optimum value of k for green belt-pollutant combination. A relationship for the estimation of this optimum value of k is not yet available.

The green belt model used here is a simple one. A more realistic model should also consider the surface features of leaves and other parts of trees which remove the pollutants from the air. For example, it is known that, if the depositing surface is wet, sticky or hairy, the deposition efficiency of pollutant on such a surface is comparatively higher. Similarly, while estimating the overall leaf area of green belt, the shadowing effect on leaf by the surrounding leaves need to be considered. While using the value of deposition velocity in the model, care should be taken to ensure that resuspension (or emission) of absorbed pollutant is accounted in the final value of deposition velocity.

The two examples given in this paper indicate that a green belt is more effective when it is planted around the source rather than around the receptors. Further, the green belt is more appropriate for near ground surface releases of pollutants. Since the normal releases of pollution are, hopefully, within permissible limits, it is the accidential releases of pollution which take place near the ground that are of major health concern. As a green belt is more effective for ground level releases, the planting of green belt around hazardous industries is an important environmental safety measure for controlling the harmful consequences of industrial disasters in public domain.

In order to effectively use the green belt for pollution attenuation, a need for further investigation exists, with some of the specific research needs as follows:

(i) Development of instrumental techniques to measure the deposited flux of pollutant on vegetative canopy in order to have a more realistic estimate of deposition velocity representative of canopy.

(ii) Conducting field studies using gaseous pollutant to evaluate the uptake rate pollution depletion rate by plants with travel distance. Such experiments have been done at Brookhaven National Laboratories, in the USA but only for pollen (i.e. particulates) using pine forests.

(iii) Development of simple and accurate methods for estimation of foliage surface area density of different trees.

(iv) Search for pollution resistant tree species having high pollution uptake rates.

(v) Development of models which estimate the pollution uptake rate for the whole canopy using the information on uptake rate of plants.

(vi) Models for biogeochemical cycling through the plant-atmosphere-soil system. This is for understanding of long term effects of absorbed pollutants on plants. Presently some information is available on cycling of radionuclides in ecosystems.

(vii) Investigations of pollutant-plant reactions at various biological levels of the plant, e.g. cellular, sub-cellular, and organism levels.

6 REFERENCES

Bache, D.H., 1979. Particulate transport within plant canopies - II. Prediction of deposition velocities. Atmospheric Environment 13, pp. 1681-1687.
_____, 1981. Analysing particulate deposition to plant canopy. Atmospheric Environment 15:1759-1761.
Bennet, J.H., and Hill, A.C., 1975. Interactions of air pollutants with canopies of vegetation. In: J.B. Mudd and T.T. Kozlowski (Editors), Response of Plants to Air Pollution. Academic Press, New York, pp. 273-304.
Bernatzky, A., 1968. Protection plantings for air purification and improvement of environmental conditions. Tree J., 2: 37-42.
Chadwick, R.C., and Chamberlain, A.C., 1970. Field loss of radionuclides from grass. Atmospheric Environment 4: 51-56.
Chamberlain, A.C., 1953. Aspects of travel and deposition of aerosol and clouds. A.E.R.E. Report H.P. 1261, Atomic Energy Research Establishment, Harwell, Berks. U.K.
_____, 1970. Interception and retention of radioactive aerosols by vegetation.

Atmospheric Environment 4: 57-78.

Flemming, G., 1967. When can forest belts reduce the emissions concentration. Luft and Kaeltetchnik 6: 255-258.

Ganguly, A.K., 1976. A proposal for a national programme of development of perennial green envelope and large scale afforestation. Scavenger, 6, 8-17, published by "SOCLEEN", Garden Resort, 606, Bombay-71, India.

Gupta, V.K. and Kapoor, R.K., 1985. Reducing the consequences of reactor accidents with a green belt. Nuclear Technology, 70, 2: 204-214.

Hanson, G.P. and Throne, L., 1970. A partial pollution solution-plant trees. Lasca Leaves, 20, pp. 35-36.

Hosker, R.P., Jr., 1973. Estimates of dry deposition and plume depletion over forest and grassland. In: The Physical Behaviour of Radioactive Contaminants in the Atmosphere, pp. 291-310. IAEA Symposium SM-181. International Atomic Energy Agency, Vienna, Austria.

_____, and Lindberg, S.E., 1982. Review: atmospheric deposition and plant assimilation of gases and paticulates. Atmospheric Environment 16: 889-910.

Kalyushnyi, Y., et al., 1952. Effectiveness of sanitary clearance zones between industrial enterprises and residential quarters. Gicienai Sanit. 4, 179.

Kapoor, R.K., 1979. A study of Height of Atmospheric Boundary Layer and Natural Ventilation for Air Pollution of some of the Indian Cities, Ph.D. Thesis, University of Bombay.

_____, and Gupta, V.K., 1984. A pollution attenuation coefficient concept for optimization of green belt. Atmospheric Environment, 18, pp. 1107-1113.

_____, 1985. Effectiveness of green belt to protect the Taj Mahal against air pollution. National Seminar on Pollution Control and Environmental Management, NEERI, Nagpur, March 17-19, 1985.

Monteith, J.L., 1975. Vegetation and the Atmosphere, Vol. 1 - Principles, Vol. II - Case Studies. Academic Press.

Mudd, J.B., and Kozlowski, T.T., 1975. Response of Plants to Air Pollution, Academic Press.

Raynor, G.S., Hayes, J.V., and Ogden, E.C., 1974. Particulate dispersion into and within forest. Boundary-Layer Met. 7: 429-456.

Sehmel, G.A., 1980. Particle and gas dry deposition: a review. Atmospheric Environment 14, pp. 983-1012.

Shinn, J.H., 1979. Problems in the assessment of air pollution effects on Vegetation. In: J.R. Pfaffin and E.N. Ziegler, (Editors), Advances in Environmental Science and Engineering, Gordon and Breach, New York, pp. 88-105.

Shirvaikar, V.V. and Abrol, V., 1978. Manual of Dose Evaluation from Atmopsheric Release, Report No. BARC/I-503. Bhabha Atomic Research Centre, Bombay, India.

Slinn, W.G.N., 1982. Predictions for particle deposition to vegetative canopies. Atmospheric Environment 16, pp. 1785-1794.

U.S. EPA, 1978. Guideline on Air Quality Models. OAQPS Guideline Series No. 1.2-080, EPA Report No. EPA-450/2-78-027 (Available from EPA Office of Air Quality Plan and Studies, Research Triangle Park, NC U.S.A.).

WASH-1400, 1975. Reactor Safety Study: An Assessment of Accident Risks in U.S. Commercial Nuclear Power Plants, Appendix VI. U.S. Nuclear Regulatory Commission, Report NUREG-75/014, National Technical Information Service.

Warren, J.L,, 1973. Green space for air pollution control. N.C. State University Sch. For. Resour. Tech. Rep. 50., Raleigh, N.C.

DISPERSION OF A REACTIVE AIR POLLUTANT IN A TWO LAYERED ENVIRONMENT: EFFECT OF GREEN BELT

J.B. SHUKLA and R.S. CHAUHAN
Department of Mathematics, Indian Institute of Technology, Kanpur, 208016, INDIA

1 ABSTRACT

The effect of variable wind velocity and diffusion coefficient with height on the dispersion of air pollutant from point and line sources has been studied by dividing the inversion layer into two parts where wind velocity and diffusion coefficient are assumed to be constant in each layer taking smaller values in lower layer in comparison to the upper. The effect of chemical reaction and dry deposition on the ground has also been taken into account. The analysis has been applied to study the reduction of concentration of air pollutant due to presence of a green belt in the wind direction.

2 INTRODUCTION

In general, the dispersion of pollutants is governed by the processes of molecular diffusion and convection. In the atmosphere, dispersion depends upon the types and number of sources, stack heights, various meteorological factors (such as wind, temperature inversion, rainout washout) and topography of the terrain (Pasquill (1962); Seinfeld (1975)). Several studies have been conducted to understand the process of pollutant dispersion by including some of the above mentioned factors (Smith (1957); Pasquill (1962); Hoffert (1972); Slinn (1974); Scriven and Fisher (1975); Lamb and Seinfeld (1975); Calder (1977); Ermak (1977); Alam and Seinfeld (1981); Reda and Carmichael (1982); Llewelyn (1983); Karamchandani and Peters (1983)). In particular, the redistribution of a gas plume caused by reversible washout has been investigated by Slinn (1974).

The modelling of atmospheric pollution by nitrogen and sulfur dioxides has been studied recently by Alam and Seinfeld (1981) and Reda and Carmichael (1982). The development of a second generation mathematical model for urban air pollution has been presented by Gregory et al. (1982). The effect of a foggy environment on reversible absorption of a pollutant from an area source has recently been studied by Shukla et al. (1982).

The transport of gases and particulate matter within plant and vegetable canopies has also been studied. Petit et al. (1976) results concern characteristics of air flow within and above a forest by calculating SO_2 fluxes at the top of the canopies. Bache (1979) used a modified form of diffusion equation to study particulate transport within and above the foliage canopy. Slinn (1982) gave a theoretical framework to predict particle deposition due to vegetation by considering a variable wind velocity profile. A review of atmospheric deposition and plant assimilation of gases and particles has been presented by Smith (1981) and Hosker et al. (1982) wherein a mathematical model for aerosol depletion and deposition on forests employs a modified form of convective diffusion equation with reaction terms. This model develops the interaction between forest structure and open field by considering forest aerodynamics and aerosol characteristics (Wiman (1985)) .

From the above studies it may be speculated that if a suitable green belt in the form of forest is provided around the pollutant source, some distance away but close to the place (e.g. a habitat, a historical monument) to be protected, then it is possible to protect the endangered area. In this direction Kapoor and Gupta (1984) studied the attenuation of an inert pollutant by a green belt under steady state conditions. In this paper, we study the dispersion of a reactive air pollutant from point and line sources. We focus on dry deposition on the ground when the wind velocity and diffusion coefficient are functions of elevation. The exact solution of the diffusion equation with a reaction term is obtained by dividing the inversion layer into two parts. Wind velocity and diffusion coefficients are taken as step functions with smaller values in the lower region than in the upper region. The effect of a green belt on the reduction of concentration of pollutant due to a removal mechanism is then discussed.

3 DISPERSION OF AIR POLLUTANT FROM A POINT SOURCE

Consider the dispersion of a reactive air pollutant from a continuous point source located at height h_s from the ground in presence of an inversion layer $(0 \leq z \leq H)$ of height H. The inversion layer is assumed to be divided into two layers: I $(0 \leq z \leq h_g)$ and II $(h_g \leq z \leq H)$, where h_g is the height of the lower layer. When the direction is taken to be in the prevalent wind direction, the steady state diffusion equations governing the concentration of the pollutant in the two regions, can be written as follows:

Region I $(0 \leq z \leq h_g)$:

$$u_1 \frac{\partial C_1}{\partial x} = K_{z1} \frac{\partial^2 C_1}{\partial z^2} + K_{y1} \frac{\partial^2 C_1}{\partial y^2} - \alpha C_1 \tag{1}$$

Boundary conditions:

(i)   at $x = 0$ $C_1 = 0$; $\tag{2}$

(ii) $C_1 = 0$ as $y \to \pm \infty$; $\tag{3}$

(iii) $\quad K_{z1} \dfrac{\partial C_1}{\partial z} = v_d\, C_1 \qquad \text{at } z = 0 \;; \hfill (4)$

(iv) $\quad K_{z1} \dfrac{\partial C_1}{\partial z} = K_{z2} \dfrac{\partial C_2}{\partial z}, \qquad C_1 = C_2 \qquad \text{at } z = h_g \;. \hfill (5)$

Region II $(h_g \le z \le H)$:

$$u_2 \frac{\partial C_2}{\partial x} = K_{z2} \frac{\partial^2 C_2}{\partial z^2} + k_{y2} \frac{\partial^2 C_2}{\partial y^2} - \alpha\, C_2 \;. \hfill (6)$$

Boundary conditions:

(i) $\quad C_2 = \dfrac{Q}{u_2}\, \delta(y)\, \delta(z - h_s) \qquad \text{at } x = 0 \;; \hfill (7)$

(ii) $\quad C_2 = 0 \qquad\quad \text{as } y \to \pm\infty \;; \hfill (8)$

(iii) $\quad \dfrac{\partial C_2}{\partial z} = 0 \qquad \text{at } z = H \;; \hfill (9)$

(iv) $\quad K_{2z} \dfrac{\partial C_2}{\partial z} = K_{z1} \dfrac{\partial C_1}{\partial z} \;; \qquad C_2 = C_1 \qquad \text{at } z = h_g \;, \hfill (10)$

where C_1, C_2 are the concentration of the pollutants in the two regions; u_i's are the mean velocity; K_{yi}, K_{zi} $(i = 1, 2)$ are diffusion coefficients; v_d is deposition velocity on the ground; α is chemical reaction rate and $\delta(\cdot)$ is Dirac delta function.

If the source lies in the first region, the boundary conditions at the source for equations (1) and (6) are

$$C_1 = \frac{Q}{u_1}\, \delta(y)\, \delta(z - h_s) \quad \text{at } x = 0 \;,$$

$$\hfill (11)$$

$$C_2 = 0 \qquad\qquad \text{at } x = 0 \;.$$

Using the following dimensionless quantities

$$\bar{x} = \frac{K_{z\,max}\, x}{u_{max} H^2}, \qquad \bar{z} = \frac{z}{H}, \qquad \bar{u}_i = \frac{u_i}{u_{max}},$$

$$\overline{K}_{zi} = \frac{K_{zi}}{K_{z\,max}} \quad, \qquad \overline{v}_d = \frac{v_d h}{K_{z\,max}} \quad, \qquad \overline{h}_g = \frac{h_g}{H} \quad,$$

$$\overline{\alpha} = \frac{k\,H^2 \alpha}{K_{z\,max}} \quad, \qquad \overline{C}_i = \frac{u_{max}\,H^2}{Q}\,C_i \quad, \tag{12}$$

the equations (1-10) become (dropping the bars for convenience)

Region I $(0 \le z < h_g)$

$$\frac{\partial C_1}{\partial x} = \beta_1 \frac{\partial^2 C_1}{\partial y^2} + \gamma_1 \frac{\partial^2 C_1}{\partial z^2} - \alpha_1\,C_1 \quad; \tag{13}$$

$$C_1 = 0 \qquad \text{at } x = 0 \quad; \tag{14}$$

$$C_1 = 0 \qquad \text{as } y \to \pm\infty \quad; \tag{15}$$

$$\frac{\partial C_1}{\partial z} = N_1 C_1 \quad; \tag{16}$$

$$K_{z1}\frac{\partial C_1}{\partial z} = K_{z2}\frac{\partial C_2}{\partial z}, \quad C_1 = C_2 \text{ at } z = h_g \quad. \tag{17}$$

Region II $(h_g \le z \le 1)$

$$\frac{\partial C_2}{\partial z} = \beta_2 \frac{\partial^2 C_2}{\partial y^2} + \gamma_2 \frac{\partial^2 C_2}{\partial z^2} - \alpha_2\,C_2 \quad; \tag{18}$$

$$C_2 = \frac{1}{u_2}\delta(y)\,\delta(z - h_s) \qquad \text{at } x = 0 \quad; \tag{19}$$

$$C_2 = 0 \qquad \text{as } y \to \pm\infty \quad; \tag{20}$$

$$\frac{\partial C_2}{\partial z} = 0 \qquad \text{at } z = 1 \quad; \tag{21}$$

$$K_{z2}\frac{\partial C_2}{\partial z} = K_{z1}\frac{\partial C_1}{\partial z}, \quad C_1 = C_2 \text{ at } z = h_g \quad, \tag{22}$$

where

$$\gamma_1 = \frac{K_{z1}}{u_1}, \quad \gamma_2 = \frac{K_{z2}}{u_2}, \quad \alpha_1 = \frac{\alpha}{u_1}, \quad \alpha_2 = \frac{\alpha}{u_2},$$

$$N_1 = \frac{V_d}{K_{z1}}, \quad \beta_1 = \frac{K_{y1}}{K_{z\,max}u_1}, \quad \beta_2 = \frac{K_{y2}}{K_{z\,max}\,u_2},$$

$$K_{z\,max} = \max(K_{z1}, K_{z2}), \quad u_{max} = \max(u_1, u_2).$$

The solutions of (13) and (18) can be obtained with the assumption $\beta_1 = \beta_2$; i.e.

$$\frac{K_{y1}}{u_1} = \frac{K_{y2}}{u_2}, \text{ as}$$

$$C_1 = \frac{e^{(-y^2/4\beta_1 x)}}{\sqrt{4\beta_1\,x\,\pi}} \sum_{n=1}^{\infty} e^{-\delta_n^2 x} R_n \frac{((N_1/a_{11n})\sin a_{11n}z + \cos a_{11n}z)}{((N_1/a_{11n})\sin a_{11n}\,h_g + \cos a_{11n}\,h_g)}$$

$$\text{(23)}$$

$$C_2 = \frac{e^{(-y^2/4\beta_2 x)}}{\sqrt{4\beta_2\,x\,\pi}} \sum_{n=1}^{\infty} e^{-\delta_n^2 x} R_n \frac{(\tan a_{12n}\sin a_{12n}\,z + \cos a_{12n}\,z)}{(\tan a_{12n}\sin a_{12n}\,h_g + \cos a_{12n}\,h_g)}$$

where

$$R^n = \frac{(\tan a_{12n}\sin a_{12n}\,h_s + \cos a_{12n}\,h_s)}{F(\tan a_{12n}\sin a_{12n}\,h_g + \cos a_{12n}\,h_g)}$$

$$F = \frac{u_1 P_1^2}{4a_{11n}} [2a_{11n}h_g(1 + (N_1^2/a_{11n}^2)$$

$$+ (1 - (N_1^2/a_{11n}^2)\sin 2a_{11n}h_g - (2N_1/a_{11n})(\cos 2a_{11n}h_g - 1)]$$

$$+ (u_2 P^2/4a_{12n})[2a_{12n}(1 - h_g)(1 + \tan^2 a_{12n})$$

$$+ (1 - \tan^2 a_{12n})(\sin(2a_{12n}) - \sin(2a_{12n}\,h_g))$$

$$- 2\tan a_{12n}(\cos 2a_{12n} - \cos 2a_{12n}\,h_g)],$$

$$a_{11n} = \sqrt{(\delta_n^2 - \alpha_1)/\gamma_2} \qquad\qquad a_{12n} = \sqrt{(\delta_n^2 - \alpha_2)/\gamma_2} \ ,$$

$$P_1 = 1/(N_1/a_{11n} \ \sin a_{11n} \ h_g + \cos a_{11n} \ h_g) \ ,$$

$$P_2 = 1/(\tan a_{12n} \ \sin a_{12n} \ h_g + \cos a_{12n} \ h_g) \ ,$$

and δ_n's are solutions of the following transcendental equation

$$\frac{K_{z1}(N_1 \cos a_{11} \ h_g - a_{11} \ \sin\!\sin a_{11} \ h_g)}{((N_1/a_{11}) \ \sin a_{11} \ h_g + \cos a_{11} \ h_g)} =$$

$$\frac{K_{z2}(a_{12} \tan a_{12}\cos a_{12} \ h_g - a_{12} \sin a_{12} \ h_g)}{(\tan a_{12} \sin a_{12} \ h_g + \cos a_{12} \ h_g)} \ . \tag{25}$$

Now, if the source lies in region I, the concentration of pollutant in the other region is given by

$$C_1 = \frac{e^{-(y^2/4\beta_1 x)}}{\sqrt{4\beta_1 \ x \ \pi}} \sum_{n=1}^{\infty} e^{-\delta_n^2} \times R_{n0} \frac{((N_1/a_{11n}) \ \sin a_{11n}z + \cos a_{11n}z)}{((N_1/a_{11n}) \ \sin a_{11n} \ h_g + \cos a_{11n} \ h_g)} \ , \tag{26}$$

$$C_2 = \frac{e^{-(y^2/4\beta_2 x)}}{\sqrt{4\beta_2 \ x \ \pi}} \sum_{n=1}^{\infty} e^{-\delta_n^2} \times R_{n0} \frac{(\tan a_{12} \ \sin a_{12n} \ z + \cos a_{12n} \ z)}{(\tan a_{12n} \sin a_{12n} \ h_g + \cos a_{12n} \ h_g)} \ , \tag{27}$$

where

$$R_{no} = \frac{N \ a_{11n}^{-1} \ \sin a_{11n} \ h_s + \cos a_{11n} \ h_s}{F(\tan a_{12n} \ \sin a_{12n} \ h_g + \cos a_{12n} \ h_g)} \ .$$

4 DISPERSION OF POLLUTANT FROM A LINE SOURCE

Consider the dispersion of reactive air pollutant from a continuous line source in the two layered inversion region as before. In this case, the partial differential equations describing the concentration of pollutant in each region are

Region I $(0 \leq z \leq h_g)$

$$u_1 \frac{\partial C_1}{\partial x} = K_{z1} \frac{\partial^2 C_1}{\partial z^2} - \alpha C_1 \tag{28}$$

with boundary conditions

(i) $C_1 = 0$, at $x = 0$; $\tag{29}$

(ii) $K_{z1} \frac{\partial C_1}{\partial z} = v_d C_1$,   at $z = 0$; $\tag{30}$

(iii) $K_{z1} \frac{\partial C_1}{\partial z} = K_{z2} \frac{\partial C_2}{\partial z}$, $C_1 = C_2$ at $z = h_g$. $\tag{31}$

Region II $(h_g \leq z \leq H)$

$$u_2 \frac{\partial C_2}{\partial x} = K_{z2} \frac{\partial^2 C_2}{\partial z^2} - \alpha C_2 \tag{32}$$

(i) $C_2 = \dfrac{Q}{u_2} \delta(z - h_g)$ at $x = 0$; $\tag{33}$

(ii) $\dfrac{\partial C_2}{\partial z} = 0$ at $z = H$; $\tag{34}$

(iii) $K_{z2} \dfrac{\partial C_2}{\partial z} = K_{z1} \dfrac{\partial C_1}{\partial z}$, $C_2 = C_1$ at $z = h_g$. $\tag{35}$

If the source lies in the first layer, the boundary conditions at the source for equations (28) and (32)

$$C_1 = \frac{Q}{u_1} \delta(z - h_s) \quad \text{at } x = 0 ;$$

$$\tag{36}$$

$$C_2 = 0 \qquad\qquad \text{at } x = 0 .$$

The solutions of equations (28) and (32) subject to the above boundary conditions can be obtained as before and one written in the dimensionless form, for each region, follows:

$$C_1 = \sum_{n=1}^{\infty} R_n \, e^{-\delta_n^2 \, x} \, \frac{((N_1/a_{11n}) \sin a_{11n} \, z + \cos a_{11n} \, z)}{((N_1/a_{11n}) \sin a_{11n} \, h_g + \cos a_{11n} \, h_g)} ; \tag{37}$$

$$C_2 = \sum_{n=1}^{\infty} R_n \, e^{-\delta_n^2 \, x} \, \frac{(\tan a_{12n} \, z + \cos a_{12n} \, z)}{(\tan a_{12n} \, \sin a_{12n} \, h_g + \cos a_{12n} \, h_g)} . \tag{38}$$

When the source lies in first region, the corresponding distributions of C_1 and C_2 are given as follows:

$$C_1 = \sum_{n=1}^{\infty} R_{n0} \, e^{-\delta_n^2 \, x} \, \frac{((N_1/a_{11n}) \sin a_{11n} \, z + \cos a_{11n} \, z)}{((N_1/a_{11n}) \sin a_{11n} \, h_g + \cos a_{11n} \, h_g)} ; \tag{39}$$

$$C_2 = \sum_{n=1}^{\infty} R_{n0} \, e^{-\delta_n^2 \, x} \, \frac{(\tan a_{12n} \, \sin a_{12n} \, z + \cos a_{12n} \, z)}{(\tan a_{12n} \, \sin a_{12n} \, h_g + \cos a_{12b} \, h_g)} \tag{40}$$

where various parameters are the same as defined in the case of point source.

To study the effects of various parameters on the concentration distribution of air pollutant, the following values of parameters are chosen by keeping in view that the wind velocity and diffusion coefficient near the grounds are smaller: $u_1 = 0.55$, $u_2 = 1.0$, $K_{z1} = 0.55$, $K_{z2} = 1.0$, $\alpha = 0.55$, $H_s = 0.1$ and $h_g = 0.05$, $\beta_1 = \beta_2 = 10.0$. The expressions for C_1, C_2 given by equations (23), (24), (37), and (38) are computed and plotted in Figures (1)-(4).

It is found that the concentration of pollutant decreases as downwind distance increases in both the cases of point and line sources. From Figures (1)-(4), it is also noted that the concentration of air pollutant decreases due to deposition on the ground as well as with chemical reaction parameter.

The effects of stepwise variation in the wind velocity and diffusion coefficient on the concentration distribution is shown in Figure 5 for $h_s = 0.1$, $h_g = 0.09$, $u_2 = 1.0$, $K_{z2} = 1.0$, $\alpha = 0.55$, $N_1 = 2.0$ and

(i) $u_1 = 0.55$ $K_{z1} = 0.55$,

(ii) $u_1 = 1.0$ $K_{z1} = 1.0$.

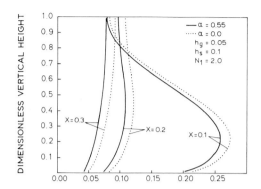

FIGURE 1 Effect of Chemical Reaction in the case of Point Source

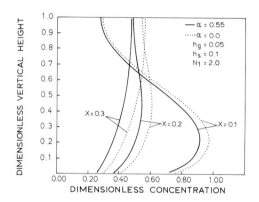

FIGURE 2   Effect of Chemical Reaction in the Case of Line Source

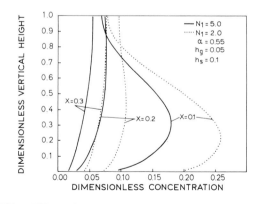

FIGURE 3 Effect of dry deposition in the case of Point Source

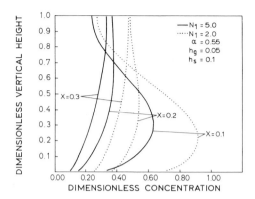

FIGURE 4 Effect of dry deposition in the case of Line Source

It is first noted from this figure that the concentration profiles for the two cases intersect at the interface of the two layers and the concentration in the lower for the $u_1 = 0.55$, $K_{z1} = 0.55$ is less than the case $u_1 = 1.0$, $K_{z1} = 1.0$. In the upper layer however the opposite result is seen.

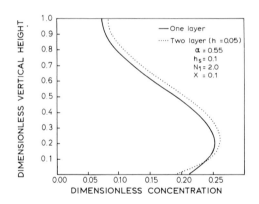

FIGURE 5 Vertical Concentration Profile when the point source located at $(0, 0, 01)$

5 EFFECT OF GREEN BELT

Consider the dispersion of a reactive air pollutant from a point source as discussed in Section 3. Assume that a green belt (a suitable tree plantation capable of absorbing a pollutant) of thickness d is located at a distance x_1 from the source along the wind direction and extending in y-direction. The physical situation is illustrated in Figure 6. It may be noted here that the inversion region is divided into four compartments with the green belt assumed located in compartment III. The partial differential equation governing the concentration distribution of the air pollutant in the regions I, II as well as their solutions have been presented in Section 2. The corresponding equations for regions III and IV follow:

Region III $(0 \leq z \leq h_g)$ $\qquad$ $x \geq x_1$

$$u_3 \frac{\partial C_3}{\partial x} = K_{y3} \frac{\partial^2 C_3}{\partial y^2} + K_{z3} \frac{\partial^2 C_3}{\partial z^2} - (k + \lambda) C_3 \qquad (41)$$

and boundary conditions are

$$C_3 = C_1 \qquad \text{at } x = x_1 ; \qquad (42)$$

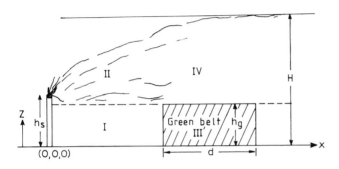

FIGURE 6 Green belt physical characteristics

$$C_3 = 0 \qquad \text{as } y \to \pm \infty ; \qquad (43)$$

$$K_{z3} \frac{\partial C_3}{\partial z} = K_{z4} \frac{\partial C_4}{\partial z} , C_3 = C_4 \qquad \text{at } z = h_g ; \qquad (44)$$

$$K_{z3} \frac{\partial C_3}{\partial z} = v_{d_3} C_3 \qquad \text{at } z = 0 . \qquad (45)$$

Region IV $(h_g \leq z \leq H)$

$$u_4 \frac{\partial C_4}{\partial x} = K_{y4} \frac{\partial^2 C_4}{\partial y^2} + K_{z4} \frac{\partial^2 C_4}{\partial z^2} - k C_4 \qquad (46)$$

and boundary conditions are

$$C_4 = C_2 \qquad \text{at } x = x_1 ; \tag{47}$$

$$C_4 = 0 \qquad \text{as } y \to \pm \infty ; \tag{48}$$

$$K_{z4} \frac{\partial C_4}{\partial z} = K_{z3} \frac{\partial C_3}{\partial z} , \; C_4 = C_3 \qquad \text{at } z = h_g ; \tag{49}$$

$$\frac{\partial C_4}{\partial z} = 0 \qquad \text{at } z = H ; \tag{50}$$

where λ is the depletion rate of pollutant due to the green belt and the other parameters are as in Sections 3 and 4. The constants C_1 and C_2 are given by equations (23), (24) at $x = x_1$ respectively.

It is convenient to cast the problem in dimensionless form. Employing dimensionless quantities given by equation (12), we have

Region III $(0 \le z \le h_g)$ $\qquad x \ge x_1$

$$\frac{\partial C_3}{\partial x} = \beta_3 \frac{\partial^2 C_3}{\partial y^2} + \gamma_3 \frac{\partial^2 C_3}{\partial z^2} - (\alpha_3 + \lambda_3) C_3 ; \tag{51}$$

$$C_3 = C_1 \qquad \text{at } x = x_1 ; \tag{52}$$

$$C_3 = 0 \qquad \text{as } y \to \pm \infty ; \tag{53}$$

$$\frac{\partial C_3}{\partial z} = N_3 C_3 \qquad \text{at } z = 0 ; \tag{54}$$

$$K_{z4} \frac{\partial C_4}{\partial z} = K_{z3} \frac{\partial C_3}{\partial z} , \; C_3 = C_4 \qquad \text{at } z = h_g . \tag{55}$$

Region IV $(h_g \le z \le 1)$ $\qquad x \ge x_1$

$$\frac{\partial C_4}{\partial x} = \beta_4 \frac{\partial^2 C_4}{\partial y^2} + \gamma_4 \frac{\partial^2 C_4}{\partial z^2} - \alpha_4 C_4 ; \tag{56}$$

$$C_4 = C_2 \qquad \text{at } x = x_1 ; \tag{57}$$

$$C_4 = 0 \qquad \text{as } y \to \pm \infty ; \tag{58}$$

$$\frac{\partial C_4}{\partial z} = 0 \qquad \text{at } z = 1 ; \tag{59}$$

$$K_{z3} \frac{\partial C_3}{\partial z} = K_{z4} \frac{\partial C_4}{\partial z} , \, C_3 = C_4 \text{ at } z = h_g , \tag{60}$$

where

$$\beta_4 = \frac{K_{y4}}{K_{z \, max} \, u_4} , \qquad \beta_3 = \frac{K_{y3}}{K_{z \, max} \, u_3} , \quad \gamma_4 = \frac{K_{z4}}{u}$$

$$\gamma_3 = \frac{K_{z3}}{u_3} , \, N_3 = \frac{v_{d3}}{K_{z3}} , \, \alpha_3 = \frac{\alpha}{u_3} , \, \alpha_4 = \frac{\alpha}{u_4} , \, \lambda_3 = \frac{\lambda H^2}{K_{z \, max} \, u_3} .$$

assuming that $(K_{y3}/u_3) = (K_{y4}/u_4) = (K_{y2}/u_2) = (K_{y1}/u_1)$, the solutions of equations (51) and (56) subjected to above boundary conditions are given as follows:

$$C_3 = \frac{e^{-(y^2 / 4\beta_3 x)}}{\sqrt{4\pi\beta_3 x}} \sum_{n=1}^{\infty} e^{-\delta'^2_n (x - x_1)} R'_n G_{1n}(z) ; \tag{61}$$

$$C_4 = \frac{e^{-(y^2 / 4\beta_4 x)}}{\sqrt{4\pi\beta_4 x}} \sum_{n=1}^{\infty} e^{-\delta'^2_n (x - x_1)} R'_n G_{2n}(z) ; \tag{62}$$

$$R'_n = \frac{\int_0^{h_g} u_3 C_1(x_1, z) G_{1n}(z) \, dz + \int_{h_g}^1 u_r C_2(x_1, z) G_{2n}(z) \, dz}{\int_0^{h_g} u_3 G^2_{1n}(z) \, dz + \int_{h_g}^1 u_4 G^2_{2n}(z) \, dz} \tag{63}$$

$$G_{1n}(z) = \frac{((N_3/a'_{11}) \sin a'_{11} z + \cos a'_{11} z)}{((N_3/a'_{11}) \sin a'_{11} h_g + \cos a'_{11} h_g)} ;$$

$$G_{2n}(z) = \frac{(\tan a'_{12} \sin a'_{12} z + \cos a'_{12} z)}{(\tan a'_{12} \sin a'_{12} h_g + \cos a'_{12} h_g)}$$

$$a'_{11} = \sqrt{(\delta'^2_h - \alpha_3 - \lambda_3)\gamma_3} , \qquad a'_{12} = \sqrt{(\delta'^2_h - \alpha_4)\gamma_4} ,$$

and δ'_n's are eigenvalues of the following equation

$$K_{z3} \frac{(N_3 \cos a'_{11} h_g - a'_{11} \sin a'_{11} h_g)}{((N_3/a'_{11})\sin a'_{11} h_g + \cos a'_{11} h_g)}$$

$$= \frac{K_{z4}(a'_{12} \tan a'_{12} \cos a'_{12} h_g - a'_{12} \sin a'_{12} h_g)}{(\tan a'_{12} \sin a'_{12} h_g + \cos a'_{12} h_g)} \tag{64}$$

It is noted that the C_1, C_2 involved in equation (63) are given by equations (37)-(38) when the source lies in the second layer (i.e. $h_s > h_g$) and by equations (39)-(40) where the source lies in the lower layer.

In the case of a line source, the distributions for C_3 and C_4 can be obtained by integrating equations (61)-(63) from $-\infty$ to ∞ with respect to y. In this case, the distribution for C_1 and C_2 in (63) would be given by (37)-(38) when $h_s > h_g$ and by equations (39)-(40) when $h_s < h_g$.

To study the depletion of pollutant due to a green belt, the following values of parameters have been chosen for computation by keeping in view that the wind velocity and diffusion coefficient near the ground and in the green belt are smaller than in the other regions: $u_1 = 0.55$, $u_2 = 1.0$, $u_3 = 0.5$, $u_4 = 1.0$, $K_{z1} = 0.55$, $K_{z2} = 1.0$, $K_{z3} = 0.5$, $K_{z4} = 1.0$, $\alpha = 0.55$, $\lambda = 4.0$ and $h_g = 0.05$. By taking various heights of the source the vertical concentration distribution of pollutant has been computed at the end of the green belt and depicted in Figures (7)-(9). From these figures it is noted that for a given location of the green belt, the concentration of pollutant in the presence of a green belt is much smaller than when the green belt is absent. Also, if we compare Figures 8 and 9, it is found that the depletion in the concentration due to a green belt is high if the height of source is less than the height of green belt.

The model presented in this paper is useful in evaluating the effect of green belt in protecting structures and monuments such as the Taj Mahal. It can also be utilized for designing a suitable green plantation for protection of habitats [See Kapoor and Gupta, this volume].

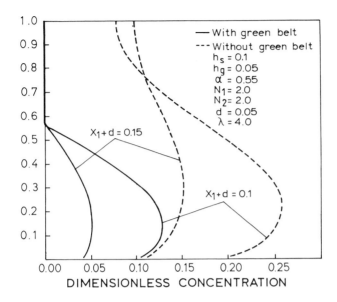

FIGURE 7 Vertical Concentration Profile when $h_s > h_g$

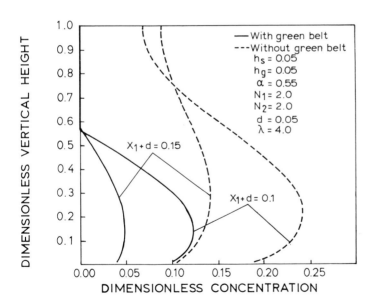

FIGURE 8 Vertical Concentration Profile when $h_s = h_g$

72

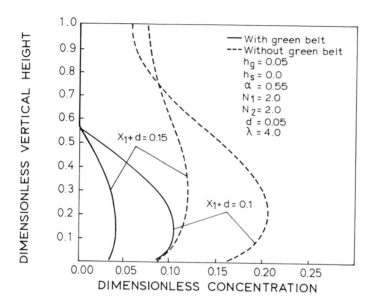

FIGURE 9 Vertical Concentration Profile when $h_s < h_g$

6 REFERENCES

Alam, M.K. and Seinfeld, J.H., 1981. Solution of the steady state, three dimensional atmospheric diffusion equation for sulphur dioxide and sulphate dispersion from point sources. Atmospheric Environment, 15: 1221-12225.

Bache, D.H., 1979a. Particle Transport within plant canopies - I: A frame work for analysis. Atmospheric Environment 13: 1257-1262.

_____, 1979b. Particulate transport within plant canopies - II: Prediction of deposition velocities. Atmospheric Environment 13: 1681-1687.

Calder, K.L., 1977. Multiple source plume models of urban air pollution: Their general structures. Atmospheric Environment 11: 403-411.

Ermak, D.L., 1977. An analytic model for air pollutant transport and deposition from a point source. Atmospheric Environment 11: 231-237.

Gotass, Y., 1982. Vertical distribution of sulphur in the atmosphere in a case of long range transport and the rate of transformation to sulphate. Atmospheric Environment 16: 1043-1046.

Gregory, J.M., Goodin, W.R., and Seinfeld, J.H., 1982. Development of a second generation mathematical model for urban air pollution. Atmospheric Environment 16: 667-696.

Hoffert, M.I., 1972. Atmospheric transport, dispersion and chemical reactions in air pollution: A review. AIAA, J. 10: 337-387.

Hosker, Jr., R.P., and Lindberg, S.E., 1982. Review: Atmospheric deposition and plant assimilation of gases and particles. Atmospheric Environment 16: 889-910.

Kapoor, R.K. and Gupta, V.K., 1984. A pollution attenuation coefficient concept for optimization of green belt. Atmospheric Environment 18: 1107-1113.

Karamchandani, P. and Peters, L.K., 1983. Analysis of the error associated with grid representation of point source. Atmospheric Environment 17: 927-933.

Lamb, R.G. and Seinfeld, J.H., 1975. Mathematical Modelling of urban air pollution: General theory. Environ. Sci. Tech. 7: 253-261.

Llewelyn, R.P., 1983. An Analytical Model for the transport dispersion and elimination of Air pollutants omitted from a point source. Atmospheric Environment 17: 249-256.

Pasquill, F., 1962. Atmospheric diffusion. Von. Nostrand., Princeton, N.J.

Petit, C., Trinite, M., and Valentin, P., 1976. Study of turbulent diffusion above and within a forest application in the case of SO_2. Atmospheric Environment 10: 1057-1063.

Reda, M., and Carmichael, C.R., 1982. Non-isothermal effects on SO_2 absorption by water droplets. Atmospheric Environment 16: 145-159.

Scheneider, T., and Grand, L., 1982. Air pollution by nitrogen oxides. Elsevier Scientific Pub. Co., N.Y.

Scriven, R.A., and Fisher, B.E.A., 1975. The long range transport of airborne material and its removal by deposition and washout. Atmospheric Environment, 9: 49-68.

Seinfeld, J.H., 1975. Air pollution: Physical and Chemical Fundamentals. McGraw Hill Book Co., N.Y.

Shukla, J.B., Nassaswamy, R., Verma, S., and Seinfeld, J.H., 1982. Reversible absorption of a pollutant from an area source in a stagnant fog layer. Atmospheric Environment 16: 1035-1037.

Slinn, W.G.N., 1974. The redistribution of gas plume caused by reversible washout. Atmospheric Environment 8: 233-239.

_____, 1976. Formulation and a solution of the diffusion deposition-resuspension problem. Atmospheric Environment 10: 763-768.

_____, 1982. Prediction for particle deposition to vegetative canopies. Atmospheric Environment, 16: 1785-1794.

Smith, W.H., 1981. Air pollution and forests. Springer-Verlag, N.Y.

Smith, F.B., 1957. The diffusion of smoke from a continuous elevated point source into a turbulent atmosphere. J. Fluid Mech. 2: 50-76.

Wiman, B.L.B., and Agren, I.G., 1985. Aerosol depletion and deposition in forests: A model analysis. Atmospheric Environment, 19: 335-367.

DISPERSON FROM A TIME DEPENDENT POINT SOURCE: APPLICATION TO
METHYL ISOCYNATE LEAKAGE IN BHOPAL, INDIA

R.S. CHAUHAN, Department of Mathematics, Indian Institute of Technology, Kanpur,
208016 INDIA
J.B. SHUKLA, Department of Mathematics, Indian Institute of Technology, Kanpur,
208016 INDIA
T.G. HALLAM, Department of Mathematics, University of Tennessee, Knoxville,
Tennessee 37996-1300 USA

1 ABSTRACT

The dispersion, from a time dependent point source, of an airborne pollutant which undergoes a first order chemical reaction forming a secondary toxic substance is studied by solving a three dimensional diffusion equation. Various examples of time dependent flux at the source are investigated to illustrate some types of chemical releases and the subsequent effects of the release. An application, which assumes that the source is located near the ground, is presented to assess the dispersion of the methyl isocynate release from the Union Carbide factory in Bhopal, India.

2 INTRODUCTION

The dispersion of air pollutant from a point source has been investigated by many (Smith (1957); Lamb and Seinfeld (1973); Dobbins (1979); Karamchandari and Peters, (1983); Llewlyn (1983)) . In general, a dispersal process depends upon meteorological conditions and removal processes such as chemical reaction, deposition, and washout (Slinn (1974); Seinfeld (1975); Novotny and Chesters (1981)) . A simple example of a chemical reaction affecting dispersion is SO_2 which can be converted to sulfate (SO_4^{2-}) aerosol in the atmosphere. Both sulfur dioxide and sulfate can be removed by wet and dry deposition. Effects of removal mechanisms on dispersal of air pollutants have been modelled primarily by considering them as first order processes (Scriven and Fisher (1975); Nordlund (1975); Sander and Seinfeld (1976); McMohan et al. (1976); Prahm et al. (1976); Sheih (1977); Calvert et al. (1978); Slinn (1980); Hales (1982)) .

Recently, Alam and Seinfeld (1981) have studied the steady state dispersion of sulfur dioxide from a continuous point source. They account for a reaction forming sulfate as a secondary pollutant and removal mechanisms such as deposition. In general, however, the sources of pollutant are time dependent and this aspect should be considered in the dispersal process. An example of a time dependent source is the leakage of methyl isocynate (MIC) gas from the Union Carbide factory at Bhopal, India. To model such a situation, a set of dynamic diffusion equations with chemical reaction can be employed. The analysis and application of a time-varying release and its dispersion are the subjects of this article.

The dispersion, from a time dependent point source, of a pollutant undergoing a first order chemical reaction and forming a secondary pollutant is investigated in two cases:

(1) Dispersion from an elevated point source where no deposition on the ground occurs.

(2) Dispersion from a point source located near the ground with deposition on the ground.

3 DISPERSION FROM AN ELEVATED POINT SOURCE

Consider the dispersion of a reactive gaseous species G from a time dependent elevated point source in the environment and forming a secondary species P. Both chemicals are assumed to be removed from the atmosphere.

The point source is taken as the origin of the coordinate system (x, y, z), where x is in the direction of prevalent wind. The concentration of both species are assumed governed by the following atmospheric diffusion equation

$$\frac{\partial C_i}{\partial t} + U \frac{\partial C_i}{\partial x} = D(\frac{\partial^2 C_i}{\partial x^2} + \frac{\partial^2 C_i}{\partial y^2} + \frac{\partial^2 C_i}{\partial z^2}) + R_i \, , i = 1, 2 . \tag{1}$$

In (1), C_1 and C_2 are concentrations of primary and secondary pollutants, $R_i(C_1, C_2)$ represents the chemical reaction term; D is the diffusion coefficient and U is the mean wind velocity. Our specific application will assume that R_i can be represented by first order processes,

$$R_1(C_1, C_2) = -kC_1 - k_g \, C_1 ,$$

$$\tag{2}$$

$$R_2(C_1, C_2) = kC_1 - k_p \, C_2 ,$$

where k is the rate at which the chemical species G is converted to P and k_g and k_p are the removal rates of G and P respectively.

The initial and boundary conditions for equation (1) can be written as

$$C_i(s, t) = 0 \qquad \text{at } t = 0 \text{ for } s = \sqrt{x^2 + y^2 + z^2} > 0 \tag{3}$$

$$C_i(s, t) = 0 \text{ as } s \to \infty \qquad t \geq 0 \qquad \tag{4}$$

$$-4\pi s^2 D \frac{\partial C_i}{\partial s} = W_i(t) \qquad \text{as } s \to 0 \qquad t \geq 0 , i = 1, 2 \tag{5}$$

The last boundary condition models the hypothesis that the point source has a prescribed time dependent flux $W_i(t)$. If there is no direct emission of secondary pollutant from the source (i.e. $W_2(t) = 0$), the boundary condition for C_2 at the source becomes

$$-4\pi s^2 D \frac{\partial C_2}{\partial s} = 0 \qquad \text{as } s \to 0, \quad t \geq 0. \tag{6}$$

The following rudimentary forms of $W_1(t)$ will be considered in the subsequent analysis:

Instantaneous flux: $\quad W_1(t) = W_0 \delta(t) \tag{7i}$

Constant flux: $\qquad W_1(t) = W \ (\text{constant}) \tag{7ii}$

Step function flux: $\quad W_1(t) = W \qquad 0 < t \leq t_0 \tag{7iii}$
$$= 0 \qquad t > t_0$$

When the removal mechanisms are represented as first order processes; that is, the functions R_i are linear in C_{ij} as in (2), the system (1)-(6) can be solved analytically in a closed form provided the form of W_1 is sufficiently simple. The forms of W_1 as hypothesized in (7) above are in this category and the solutions are presented in Appendix A. The concentration distributions are given in a dimensionless form for each representation of W_1.

To illustrate the dependence of concentration gradients on W_1, the expressions for the concentrations C_{1j}, C_{2j} $(j = 1, 2, 3)$ for the choices (7) of W_j along the central line $(x, 0, 0)$ are computed and their distributions are shown in Figures (1)-(3) for the set of parameters: $U = 4 \text{m s}^{-1}$, $D = 4 \text{ m}^2 \text{s}^{-1}$, $\alpha = .0001$, $\alpha_g = 0.0002$ and $\alpha_p = .0005$. It is observed that the concentrations C_{1j}, C_{2j} $(j = 1, 2, 3)$ decrease as downwind distance increases. It is also noted that as the dimensionless rate of chemical conversion α increases the concentration C_{1i} decreases but C_{2i} increases at a particular time and location.

When the flux is instantaneous at the source, the concentration distributions of C_{11}, C_{21} are shown in Figure 1 for different values of t (dimensionless $t = 240$ corresponds to 1 min), $W_0 = 1.0$, $\alpha = .0001$, $\alpha_g = 0.0$, 0.0002 and $\alpha_p = 0.0$, 0.0005 It is noted from this figure that as time increases both C_{11} and C_{21} decrease and the maximum in the concentration distance profile of these species moves away from the source.

For the case of constant flux, the central line concentrations C_{21} and C_{22} are depicted in Figure 2 for the values of t, α, α_g, α_p as above. In this case both C_{21} and C_{22} increase as time increases and eventually tend to steady state values.

When the flux is taken as a step function, the central line concentrations are plotted in Figure 3. The concentrations C_{13} and C_{23} decrease as time t increases for $t > t_0$. However for $t \leq t_0$ the profiles for C_{13} and C_{23} are similar to the case of constant flux.

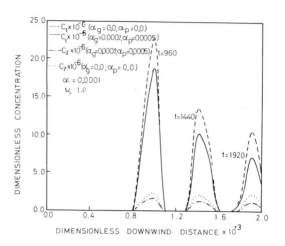

FIGURE 1 Flux is Instantaneous at the Source

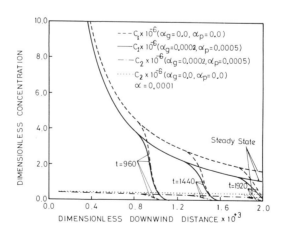

FIGURE 2 Flux is Constant at the Source

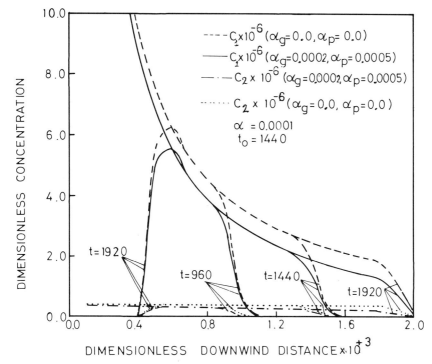

FIGURE 3 Flux is Step Type function at the source

The effect of the removal process on the concentrations C_{1i} and C_{2i} can be evaluated by comparison of the figures. It is observed that both C_{1i} and C_{2i} decrease as α_g, α_p increase.

4 DISPERSION FROM A POINT SOURCE NEAR THE GROUND

We now discuss the problem of the dispersion of a reactive gaseous pollutant from a time dependent point source located at height h_s from the ground. The pollutant is assumed to form a secondary pollutant in the atmosphere.

The unsteady state diffusion equation governing the concentrations C_1, C_2 of both species can be written as (Alam and Seinfeld (1981))

$$\frac{\partial C_i}{\partial t} + U\frac{\partial C_i}{\partial x} = D_y\frac{\partial^2 C_i}{\partial y^2} + D_z\frac{\partial^2 C_i}{\partial z^2} + R_i(C_1, C_2) \qquad i = 1, 2 \qquad (8)$$

where x is taken as the dominant wind direction, z measures the height, U is the mean wind velocity and D_y, D_z are diffusivities in y, z directions respectively. The reaction terms R_i are again given by first order processes; see equation (2).

The initial and boundary conditions for (8) are

$$C_i(x, y, z, t) = 0 \qquad\qquad\qquad t = 0 \qquad\qquad (9)$$

$$C_i(x, y, z, t) = \frac{W_i(t)}{U}\, \delta(x)\, \delta(z - h_s) \qquad \text{at } x = 0 \qquad (10)$$

$$C_i(x, y, z, t) = 0 \qquad\qquad \text{as } y \to \pm\infty \qquad (11)$$

$$D_z \frac{\partial C_i}{\partial z} = v_{d_i} C_i \qquad\qquad \text{at } z = 0 \qquad (12)$$

where $\delta(\cdot)$ represents the Dirac delta function and v_{d_i}'s are deposition velocities of the primary and secondary pollutant on the ground.

If there exists an inversion layer at height H which inhibits vertical mixing, we have

$$\frac{\partial C_i}{\partial z} = 0 \qquad \text{at } z = H \qquad\qquad (13)$$

The boundary condition (13) implies that at the point source the time dependent concentration has been prescribed in terms of flux $W_i(t)$ $(i = 1, 2, 3)$. If there is no direct emission of secondary pollutant from the source, $W_2 = 0$ or $C_2 = 0$, at the source.

For the same forms of $W_1(t)$ as in section 2 (see equation 7), the solutions of the system (8)-(13) are given in Appendix B in dimensionless form. The effect of various parameters on the concentrations C_{1j} and C_{2j} along the central line $(0, 0, h_s)$ is illustrated in Figures 4-9. The concentrations of C_{1j} and C_{2j} $(j = 1, 2, 3)$ decrease as downwind distance increases. As α' increases, the concentration of C_{1j} decreases but C_{2j} increases at a particular time and location. As expected, the effect of removal is to decrease the concentration of both species.

When the source strength is instantaneous the concentration distributions of C_{11} and C_{21} are shown in Figures (4) and (5) for different values of t, $W_0 = 1.0$, $\alpha' = 2.77$, $\alpha'_g = 0.0$, 2.00 and $\alpha'_p = 0.0, 20.0$ (for SO_2 and SO_4^{2-}). It is noted that as time increases both C_{11}, C_{21} decrease. The maximum in the concentration distance profile moves away from the source with the concentration of C_{21} being less than C_{11} for the set of parameters shown in Figures 4 and 5.

When the flux is constant, the central line concentration of both species are plotted in Figures 6 and 7 for different values of t, $\alpha = 2.77$, $\alpha_g = 0.0, 2.0$ and $\alpha_p = 0.0, 20.0$. From these figures, we observe that the concentration of each species, C_{12} and C_{22}, increases as time increases and they approach steady state values.

For $t < x$, the concentration of C and C_p are zero; that is, the pollutant front has not yet reached the point x.

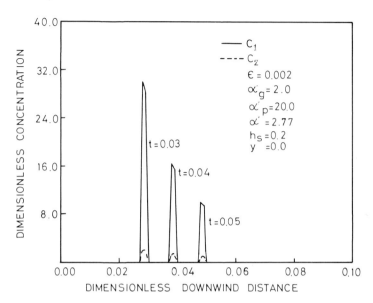

FIGURE 4 Concentration is Instantaneous at the source

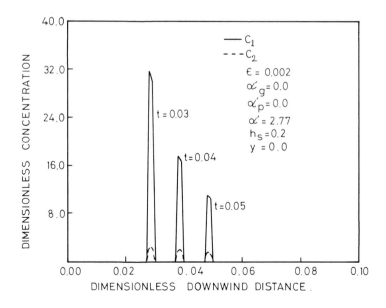

FIGURE 5 Concentration is Instaneous at the source

When the flux is taken as the step type function, the central line concentrations of C_{13} and C_{23} decrease as time increases for $t > t_0$ (see Figures 8 and 9) .

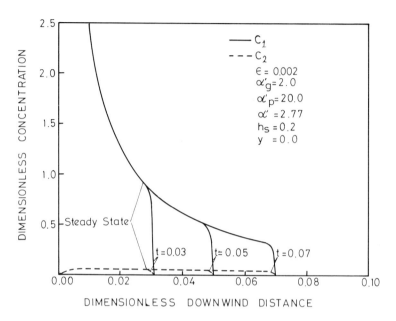

FIGURE 6 Concentration is Constant at the source

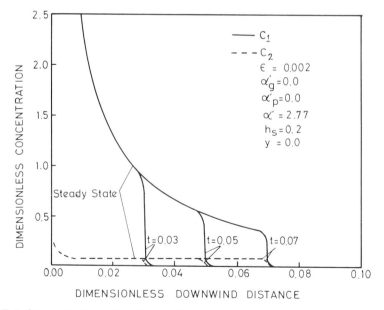

FIGURE 7 Concentration is Constant at the source

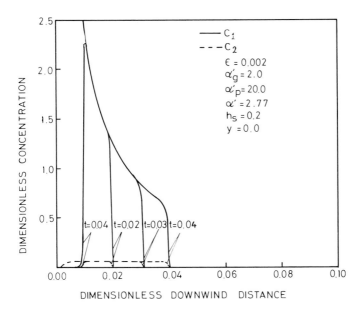

FIGURE 8   Concentration is Step type Function at the source

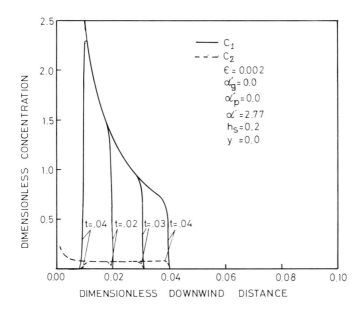

FIGURE 9 Concentration is Step Type Function at the source

5 APPLICATION TO MIC LEAKAGE IN BHOPAL, INDIA

The world is well aware of the problem of methyl isocynate (MIC) leakage from the Union Carbide factory in Bhopal, on the night of December 3, 1984. At the time of the MIC leakage a light 12 km/hour breeze was blowing in a south southeasterly direction and temperature was around 12°-14°C . Approximately 40 tons of MIC leaked out for about 1 1/2 hours. This gas, as it is heavier than air, settled over a considerable area of the city as the wind diminished. The limit of the region where serious exposure occurred was about 8-9 kms down wind from the factory. It was suggested that the gas eventually travelled over 15 kms, covering an area of over 50 square kms, affecting nearly two lakh people, thus, impacting one fourth of the city's population. A report (1984) indicates that the leakage was from the top of the 33 meter high ventline.

MIC is strongly toxic and highly unstable. It decomposes readily in the presence of moisture forming methyl amine according to the following reaction:

$$CH_3NCO \quad + H_2O \rightarrow CH_3N = C\begin{smallmatrix} \nearrow OH \\ \searrow OH \end{smallmatrix} \quad \rightarrow \quad CH_3NH_2 \quad + CO_2$$

(methyl isocynate)   (methyl amine)

A person exposed to MIC gas at a given level for about 5 minutes experiences the following kinds of effects:

0.4 ppm	~	very little effect
2.00 ppm	~	lachrymation, irritation in nose and throat ;
4.0 ppm	~	symptoms of irritation more marked ;
21.0 ppm	~	unbearable irritation of eyes, nose, and throat .

MIC destroys proteins and lipid in the lungs resulting in changes in the permeability of lung membranes and ultimately death.

The analysis and result of case (iii), Section 3 can be applied to study this situation. As the compete set of data is not available for MIC, the set of parameters in Table 1 are chosen in the computation. Since the MIC leaked under pressure, we have taken the effective source height as 40 meters. Since MIC is a highly unstable and reactive gas, the value of parameters such as a', v_{d_i} are taken greater than those associated with SO_2 (see Table 1).

TABLE 1

MIC Leakage Parameter Values

Dimensional Parameters			Corresponding non-dimensional parameters
Parameters	Values (Ref.) for SO_2	Values chosen for MIC dispersed	
k	0.05 hr^{-1} ⎤	$\simeq$ 10^{-4} sec	α' = 20.0
k_g	0.10^{-5} sec^{-1} ⎟ Alam	0.0	α'_g = 0.0
k_p	0.10^{-4} sec^{-1} ⎟ and Seinfeld	0.0	α'_p = 0.0
v_d	1 cm/sec ⎟ (1981)	2 cm/sec	N_1 = 4.0
K_v	5 m^2/sec ⎦	5 m^2/sec	
$\dfrac{K_v}{K_z}$ = ß	10	10	
u		12 km/hr	
h_s		40 meters	h_s = .04
t		1.5 hour	t = .03
x		6.60 km	x = .01
C		2 ppm	C = 1.446

Using the above data the graph of C_{13} is shown in Figure 10. It is noted from this graph for t = 0.02 (i.e., one hour) that the gas front with zero concentration has travelled about 13 kms (dimensionless x = 0.02) from the source in the wind direction. It is also seen from this graph that the line of dimensionless concentration C_{13} = 1.446 (C_{13} = 2 ppm , the level at which MIC starts affecting human beings) intersects the concentration profile for t = .02 (i.e. one hour) at a point which corresponds to a dimensionless distance, 0.01435. This is equivalent to about 10 kms in the downwind direction, approximately the limit of the region where people were seriously affected. Thus, the analysis presented in section 3, though approximate (wind velocity and diffusivities are taken as constants), shows promise as an impact assessment tool to determine dispersal of MIC leakage in Bhopal, India.

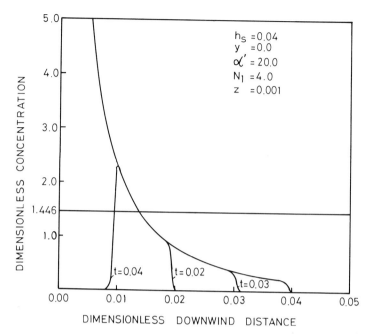

FIGURE 10 Graph of C_{13}

6 APPENDIX A

To obtain an analytical solution of equation (1), we restrict $R_i(C_1, C_2)$ by assuming that the conversion of G to P and the removal mechanisms are first order processes, $R_i(C_1, C_2)$ (i = 1, 2) that take the form (2):

$$R_1(C_1, C_2) = -k\,C_1 - k_g\,C_1$$

$$R_2(C_1, C_2) = k\,C_1 - k_p\,C_2$$

(1-A)

where k is the constant rate of conversion of G to P and k_g, k_p are their removal rates respectively.

Using the following dimensionless variables,

$$t^* = \frac{U^2 t}{D}\ ,\ x^* = \frac{U}{D}x\ ,\ y^* = \frac{U}{D}y\ ,\ z^* = \frac{U}{D}z\ ,$$

$$s^* = \frac{U}{D}s\ ,\ C_1^* = \frac{D^2}{WU}C_1\ ,\ C_2^* = \frac{D^2}{WU}C_2\ ,$$

(2-A)

$$W_0^* = \frac{W_0\,D}{WU^2}\ ,\ W^* = \frac{W_1}{W}\ ,$$

the equations for C_1 and C_2 can be written as (dropping the stars for convenience)

$$\frac{\partial C_1}{\partial t} + \frac{\partial C_1}{\partial x} = \left(\frac{\partial^2 C_1}{\partial x^2} + \frac{\partial^2 C_1}{\partial y^2} + \frac{\partial^2 C_1}{\partial z^2}\right) - (\alpha + \alpha_g)C_1 \ , \tag{3-A}$$

$$\frac{\partial C_2}{\partial t} + \frac{\partial C_2}{\partial x} = \left(\frac{\partial^2 C_2}{\partial x^2} + \frac{\partial^2 C_2}{\partial y^2} + \frac{\partial^2 C_2}{\partial z^2}\right) + \alpha C_1 - \alpha_p C_2 \ ; \tag{4-A}$$

where α, α_g and α_p are the dimensionless analogues of k, k_g, and k_p respectively:

$$\alpha = \frac{D}{U^2} k \ , \ \alpha_g = \frac{D}{U^2} k_g \ , \ \alpha_p = \frac{D}{U^2} k_p \ .$$

The initial and boundary conditions become

$$C_i(s, t) = 0 \qquad\qquad t = 0 \quad s > 0 \ , \tag{5-A}$$

$$C_i(s, t) = 0 \qquad\qquad s \to \infty \quad t \geq 0 \ , \tag{6-A}$$

$$-4\pi \, s^2 \frac{\partial C_1}{\partial s} = W_1(t) \qquad\qquad s \to 0 \quad t \geq 0 \ , \tag{7-A}$$

$$-4\pi \, s^2 \frac{\partial C_2}{\partial s} = 0 \qquad\qquad s \to 0 \quad t \geq 0 \ , \tag{8-A}$$

The forms of $W_1(t)$ are

(i) $W_1(t) = W_0 \, \delta(t)$
(ii) $W_1(t) = 1$ $\qquad\qquad\qquad\qquad\qquad\qquad\qquad$ (9-A)
(iii) $W_1(t) = 1 \qquad\quad 0 < t \leq t_0$
$\qquad\quad = 0 \qquad\quad t > t_0$

The solution of equations (3-A) and (4-A) with boundary conditions (5-A) - (8-A) for the various forms of $W_1(t)$ can be obtained as follows:

(i) Instantaneous flux, $W_1(t) = W_0 \, \delta(t)$

$$C_{11} = \frac{Q_0}{(4\pi t)^{3/2}} \exp\left[\frac{x}{2} - \left(\frac{1}{4} + \alpha + \alpha_g\right)t - \frac{s^2}{4t}\right] \tag{10-A}$$

$$C_{22} = \frac{\alpha}{\alpha_p - \alpha - \alpha_g} \ [C_{11} - \frac{Q_0}{(4\pi t)^{3/2}} \ \exp[\frac{x}{2} - (\frac{1}{4} + \alpha_p)t - \frac{s^2}{4t}]] \tag{11-A}$$

(ii) Constant flux: $W_1(t) = W$

$$C_{12} = \frac{\exp(x/2)}{4\pi s} \ \{\exp(-b^{1/2} \ s) - \frac{1}{\pi} \int_b^\infty \frac{1}{u} \ e^{-ut} \ \sin(u - b)^{1/2} \ s \ du] \tag{12-A}$$

$$C_{22} = \frac{\alpha}{\alpha_p - \alpha - \alpha_g} \ [C_{11} - \frac{e^{x/2}}{4\pi s} \ \{\exp(-b_1^{1/2} \ s) - \frac{1}{\pi} \int_{b_1}^\infty \frac{1}{u} \ e^{-ut} \ \sin(u - b_1)^{1/2} \ s \ du\}] \tag{13-A}$$

It is noted that the above expressions (12-A) and (13-A) for C_{12} and C_{22} can be obtained by integrating the corresponding distributions given by equations (10-A) and (11-A) between 0 to t.

(iii) Step Function Flux: $W(t) = W \qquad 0 < t \le t$
$$= 0 \qquad t > t_0$$

$$C_{13} = \frac{\exp(x/2)}{4\pi s} \ \{\exp(-b^{1/2} \ s)[1 - H(t - t_0)]$$

$$- \frac{1}{\pi} \int_b^\infty \frac{1}{u} \ e^{-ut} \ \sin(u - b)^{1/2} \ s[1 - e^{-ut_0} \ H(t - t_0)] \ du\} \tag{14-A}$$

$$C_{23} = \frac{\alpha}{\alpha_p - \alpha - \alpha_g} \ [C_{13} - \frac{\exp(x/2)}{4\pi s} \ \{\exp(b_1^{1/2} \ s) \ [1 - H(t - t_0)]$$

$$- \frac{1}{\pi} \int_{b_1}^\infty \frac{1}{u} \ e^{-ut} \ \sin(u - b_1)^{1/2} \ s[1 - e^{-ut_0} \ H(t - t_0)] \ du\}] \tag{15-A}$$

where $b = \frac{1}{4} + \alpha + \alpha_g$, $b_1 = \frac{1}{4} + \alpha_p$, and $H(t - t_0)$ is the Heaviside function defined by Carslaw and Jaeger (1941):

$$H(t - t_0) = 0 \qquad t \le t_0$$

$$= \frac{t}{t_0 + \varepsilon} \qquad t_0 < t \le t_0 + \varepsilon$$

$$= 1 \qquad t > t_0 + \varepsilon \tag{16-A}$$

where ε can be made arbitrarily small.

The equations (14-A) and (15-A) can also be obtained by the following:

$$C_{13} = \int_0^t C_{11}(x, y, z, t') \, dt' - H(t - t_0) \int_0^{t-t_0} C_{11}(x, y, z, t') \, dt' \qquad (17\text{-}A)$$

$$C_{23} = \int_0^t C_{21}(x, y, z, t') \, dt' - H(t - t_0) \int_0^{t-t_0} C_{21}(x, y, z, t') \, dt' \qquad (18\text{-}A)$$

7 APPENDIX B

Using the dimensionless variables,

$$t^* = \frac{D_z t}{H^2} \,, \; x^* = \frac{D_z x}{UH^2} \,, \; y^* = \frac{y}{H} \,, \; z^* = \frac{z}{H} \,, \; W_1^* = \frac{W_1}{W} \,, \; W_0^* = \frac{W_0 H^2}{WD_z} \,, \; C_i^* = \frac{UH^2 C_i}{W} \qquad (1\text{-}B)$$

the equations for C_1 and C_2 can be written in the dimensionless form as (dropping the stars for convenience)

$$\frac{\partial C_1}{\partial t} + \frac{\partial C_1}{\partial x} = \beta \frac{\partial^2 C_1}{\partial y^2} + \frac{\partial^2 C_1}{\partial z^2} - (\alpha' + \alpha'_g) C_1 \qquad (2\text{-}B)$$

$$\frac{\partial C_2}{\partial t} + \frac{\partial C_2}{\partial x} = \beta \frac{\partial^2 C_2}{\partial y^2} + \frac{\partial^2 C_2}{\partial z^2} + \alpha' C_1 - \alpha'_p C_2 \qquad (3\text{-}B)$$

where

$$\beta = \frac{D_y}{D_z} \,, \; \alpha' = \frac{kH^2}{D_z} \,, \; \alpha'_g = \frac{k_g H^2}{D_z} \,, \; \alpha'_p = \frac{k_p H^2}{D_z} \,, \; N_i = \frac{v_{d_i} H}{D_z} \,, \; i = 1, 2 \,.$$

The initial and boundary conditions take the following form

$$C_i(x, y, z, t) = 0 \qquad \text{at } t = 0 \qquad (4\text{-}B)$$

$$C_i(x, y, z, t) = 0 \qquad \text{at } y \to \pm\infty \qquad (5\text{-}B)$$

$$\frac{\partial C_i}{\partial z} = 0 \qquad \text{at } z = 1 \qquad (6\text{-}B)$$

$$\frac{\partial C_i}{\partial z} = N_i C_i \qquad \text{at } z = 0 \qquad (7\text{-}B)$$

$$C_1 = \frac{W_1(t)}{W} \delta(y) \, \delta(z - h_s) \qquad \text{at } x = 0 \qquad (8\text{-}B)$$

$C_1 = 0$

(i) $W_1(t) = W_0\, \delta(t)$

(ii) $W_1(t) = 1$ $\qquad\qquad\qquad\qquad\qquad\qquad\qquad$ (9-B)

(iii) $W_1(t) = 1$ $\qquad\qquad\qquad 0 < t \leq t_0$

$\qquad\quad\ = 0$ $\qquad\qquad\qquad\quad\ t > t_0$

The solution of equations (2-B) and (3-B) with conditions (4-B) - (9-B) can be written as follows (Shukla and Chauhan (1986)):

(i) Instantaneous flux, $W_1 = W_0\, \delta(t)$

$$C_{11}(x, y, z, t) = W_0\, P(x, y, z)\, \delta(t - x) \qquad\qquad\qquad\qquad (10\text{-}B)$$

$$C_{21}(x, y, z, t) = w_0\, Q(x, y, z)\, \delta(t - x) + \frac{\alpha'\, C_{11}\,(x, y, z, t)}{\alpha'_p - \alpha' - \alpha'_g} \qquad\qquad (11\text{-}B)$$

where

$$P(x, y, z) = e^{-(y^2/4\beta x)}/\sqrt{4\pi\beta x}\ \sum_{n=1}^{\infty} \frac{1}{P_n}\, \cos \lambda_n(h_s - 1)\, \cos \lambda_n(z - 1)\, e^{-(\alpha'_g + \alpha' + \lambda_n^2)x}$$

$$Q(x, y, z) = e^{-(y^2/4\beta x)}/\sqrt{4\pi\beta x}\ \sum_{n=1}^{\infty} \alpha' \cos \mu_n(z - 1)/(\alpha'_p - \alpha' - \alpha'_g) M_n$$

$$\{(N_1 - N_2)\cos \mu_n \sum_{m=1}^{\infty} \frac{1}{P_m}\, \cos \lambda_m \cos \lambda_m(h_s - 1)$$

$$e^{-(\alpha'_p + \mu_n^2)x} \int_0^x e^{-(\alpha' + \alpha'_g - \alpha'_p + \lambda_m^2 - \mu_n^2)\, x'} dx'$$

$$-e^{-(\alpha'_p + \mu_n^2)x} \cos \mu_n(h_s - 1)\}$$

$\lambda_n \tan \lambda_n = N_1$

$P_n = \int_0^1 \cos^2 \lambda_n(z - 1)\, dz$

$\mu_n \tan \mu_n = N_2$

$M_n = \int_0^1 \cos^2 \mu_n(z - 1)\, dz$

and the dirac delta function is defined by (Carslaw and Jaeger (1941)):

$$\delta(t) = 0 \qquad t \leq 0$$
$$= 1/\varepsilon \qquad 0 < t \leq \varepsilon$$
$$= 0 \qquad t > \varepsilon$$

(ii) Constant flux, $W(t) = W$

$$C_{12} = P(x, y, z)\, H(t - x) \qquad\qquad (12\text{-B})$$

$$C_{22} = Q(x, y, z)\, H(t - x) + \frac{\alpha'\, C_{12}\,(x, y, z, t)}{(\alpha'_p - \alpha' - \alpha'_g)} \qquad\qquad (13\text{-B})$$

where $H(t - x)$ is Heaviside function defined by (Carslaw and Jaeger (1941)):

$$H(t - x) = 0 \qquad t \leq x$$
$$= t/(x + \varepsilon) \qquad x < t \leq x + \varepsilon$$
$$= 1 \qquad t > x + \varepsilon$$

It may be noted that ε is the time in which concentration of pollutant reaches steady state at a particular location. Letting $t \to \infty$ in the above leads to the same expression for C_{12} and C_{22} obtained in the steady state analysis by Alam and Seinfeld (1981).

(iii) Step Function flux: $W(t) = W \qquad 0 \leq t \leq t_0$
$$= 0 \qquad t > t_0$$

$$C_{13}(x, y, z, t) = P(x, y, z)\, [H(t - x) - H(t - t_0 - x)\, H(t - t_0)] \qquad\qquad (14\text{-B})$$

$$C_{23}(x, y, , t) = B(x, y, z, t) + \frac{\alpha'\, C_{13}\,(x, y, z, t)}{(\alpha'_p - \alpha' - \alpha_g)} \qquad\qquad (15\text{-B})$$

$$B(x, y, z, t) = Q(x, y, z)\, [H(t - x) - H(t - t_0 - x)\, H(t - t_0)] \ .$$

8 REFERENCES

Alam, M.K. and Seinfeld, J.H., 1981. Solution of steady state, three dimensional atmospheric diffusion equation for sulfur dioxide and sulfate dispersion from point sources. Atmospheric Environment 15: 1221-1225.

Astarita, G., Wei, J., and Iorio, G., 1979. Theory of Dispersion, transformation, and deposition of atmospheric pollutant using modified Green's functions. Atmospheric Environment 13: 239-246.

Calvert, J.G., Su, F., Bottenheim, J.W., and Strausz, O.P., 1978. Mechanism of the homogeneous oxidation of sulfur dioxide in the troposphere. Atmospheric Environment 12: 197-226.

Carslaw, H.S., and Jaeger, H.C., 1941. Operational methods in applied mathematics. Dover Publication Inc., New York.

Dobbins, R.A., 1979. Atmospheric Motion and Air Pollution, John Wiley and Sons, New York.

Ermark, D.L., 1977. An analytical model for air pollutant transport and deposition from a point source. Atmospheric Environment 11: 231-237.

Hales, J.M., 1982. SMICK - A scavenging model incorporating chemical kinetics. Atmospheric Environment 16: 1717-1724.

Karamchandani, P. and Peters, L.K., 1983. Analysis of the error associated with grid representation of point source. Atmospheric Environment 17: 927-933.

Lamb, R.G. and Seinfeld, J.H., 1973. Mathematical modelling of urban air pollution: General theory. Environ. Sci. Tech. 7: 253-261.

Llewelyn, R.P., 1983. An analytical model for the transport dispersion and elimination of air pollutants emitted from a point source. Atmospheric Environment 17: 249-256.

McMohan, T.A., Denison, P.J. and Fleming, R., 1976. A long-distance air pollution transport model incorporating washout and dry deposition components. J. Appl. Met. 7: 160-167.

Nordlung, G.G., 1975. A quasi-Lagrangian cell method for calculating long-distance transport of atmospheric pollution. J. Appl. Met. 14: 1095-1104.

Novotony, V., and Chesters, G., 1981. Handbook of Nonpoint Pollution. Van Nostrand Reinhold Company, New York.

Praham, L.P., Torp, U., and Stern, R.M., 1976. Deposition and transformation rate of sulfur oxides during atmospheric transport over the Atlantic. Tellus 28: 355-372.

Report, 1984. A reconstruction of the gas tragedy, Eklavya Publication, El/208, Arora Colony, Bhopal, India.

Sander, S.P. and Seinfeld, J.H., 1976. Chemical kinetics of the homogeneous atmospheric oxidations of sulfur dioxide. Environ. Sci. Tech. 10: 1114-1123.

Scriven, R.A., and Fisher, B.E.A., 1975. The long-range transport of air borne material and its removal by deposition and washout. Atmospheric Environment 9: 49-68.

Seinfeld, J.H., 1975. Air pollution: Pyysical chemical fundamentals. McGraw-Hill Co., New York.

Sheih, C.M., 1977. Application of a statistical trajectory model to the simulation of sulfur pollution over north eastern United States. Atmospheric Environment 11: 173-178.

Shukla, J.B. and Chauhan,R.S., 1986. Dispersion from time dependent point source. Atmospheric Environment, to appear.

Slinn, W.G.N., 1974. The redistribution of a gas plume caused by reversible washout. Atmospheric Environment 8: 233-239.

_____, 1980. Precipitation scavenging. In Atmospheric Science and Power Production - 1980. (Edited by D. Randerson) U.S. Dept. of Energy, Washington, D.C.

Smith, F.B., 1957. The diffusion of smoke from a continuous elevated point source into a turbulent atmosphere. J. Fluid Mech., 2: 50-76.

PART II - WATER POLLUTION

TAKING ADVANTAGE OF TOPOGRAPHY IN THE SITING OF DISCHARGE IN RIVERS

RONALD SMITH
Department of Applied Mathematics and Theoretical Physics, University of Cambridge, Siler Street, Cambridge CB3 9EW.

1 ABSTRACT

Rivers provide a cheap and convenient way of disposing of industrial, agricultural and domestic waste products. At extremely large distances downstream (several hundred times the river width) it is only the total contamininent load that is important. However, at more modest distances, the siting of the discharge can have a major influence. For example, in a wide river with uniformly sloping sides, a one-third increase in the offshore distance of a steady discharge reduces the peak concentration at the shoreline by a factor of two. Topographic features, such as sand banks, can make the sensitivity to discharge location even more dramatic. This paper reviews recent work on choosing discharge sites, for continuous and for sudden releases of contaminants. Particular attention is given to simple means of identifying good sites.

2 INTRODUCTION

Typical industrial usages of water are 600 tonnes of water for every tonne of nitrate fertilizer manufactured, 300 tonnes for every tonne of steel and 250 tonnes for every tonne of paper. Such water is used for washing, cooling, and the conveyance of other substances either in solution or in suspension. The volume of water, as compared with other "raw materials" makes the availability of large quantities of water a prime consideration in the siting of new factories.

When it emerges from the factory the water is polluted. In lakes or tidal water there is the disadvantage that the polluted water can return and get repeated dosages of contamination. Rivers are more desirable, in that the flow continually brings fresh water and sweeps away the contaminant. Of course, this desirability of rivers tends to bring more than just a single major user. As the number of intakes and outlets along a river increases, so does the scope for serious pollution. Indeed, in England and Wales major legislation was required as long ago as 1876 with the Rivers (Prevention of Pollution) Act.

The obvious way of controlling pollution is to limit the total loading imposed upon the river. However, it takes a downstream distance of the order of 100 times the river breadth for a discharge to become well-mixed across a river. This would be 10 kilometres for a 100m wide river. Thus, there is plenty of scope for local pollution even when the bulk standards are satisfied. This local behaviour is dependent upon the siting of the discharge, and is the subject of this review.

3 TRANSVERSE DIFFUSION EQUATION

It only takes a downsteam distance of the order of 40 water depths for a contaminant to become vertically well-mixed (Smith (1979)). This is so short compared with the distances for mixing across a river, that in practice the contaminant can be regarded as being vertically well-mixed. For a steady flow the appropriate form of the advection-diffusion equation is

$$h\partial_t c + h u \cdot \nabla c = \nabla \cdot (h\kappa \cdot \nabla c) , \text{ with } \nabla \cdot (hu) = 0 \qquad (3.1)$$

where h is the water depth, u the steady flow velocity, c the concentration, and κ the horizontal contaminant-dispersion tensor (which is assumed to incorporate the effects of vertical shear as well as turbulence).

A second major simplification is that the containment plume is greatly elongated in the flow direction. (A photograph to illustrate this feature is given in Figure 5.5 of Fischer, List, Koh, Imberger & Brooks, 1979). Physically this can be attributed to the disparity between the bulk velocity u and the much weaker turbulent velocities u_* associated with the mixing process. Thus, diffusion is only important across the flow. In a straight channel this leads to the equation

$$h\partial_t c + h u \partial_x c = \partial_y (h\kappa_{22} \partial_y c) , \qquad (3.2)$$

where x and y are the longitudinal and transverse coordinates.

In a meandering stream it is first necessary to use a coordinate system aligned along and across the depth-averaged flow (Figure 1). The generalization of (3.2) is then

$$m_1 m_2 h\partial_t c + m_2 h u \partial_x c = \partial_y ((m_1/m_2) h \kappa_{22} \partial_y c) , \qquad (3.3a)$$

with

$$\partial_x (m_2 h u) = 0 , \qquad (3.3b)$$

where m_1, m_2 are the metric coefficients (Yotsukura & Sayre 1976). The river banks are streamlines. Thus, the no-flux boundary conditions for equation (3.3a) are applied on lines of constant y :

$$(m_1/m_2 h \kappa_{22} \partial_y c = 0 \quad \text{on } y = y_R , y_L . \qquad (3.4)$$

The use of the flow-following coordinates is of particular importance in numerical schemes (Fischer 1969) because otherwise numerical diffusion associated with the advection can dominate the true cross-stream diffusion.

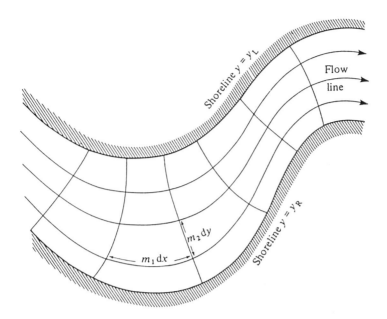

FIGURE 1 Flow-following coordinate system for a meandering channel.

4 EXACT SOLUTIONS FOR STEADY DISCHARGES

Rivers are rarely straight, and never infinitely long. However, the physical insight gained from studying this idealism justifies the considerable effort that has been put into the solution of equation (3.2).

In other physical contexts the advection-diffusion equation occurs without the presence of the depth $h(y)$. Exact solutions in terms of Bessel functions for power-law velocities and diffusivities

$$u = u_0(y/y_0)^r , \quad \kappa_{22} = \kappa_0(y/y_0)^s ,$$ (4.1)

have been presented by Lauwerier (1954) and by F.B. Smith (1957). The quantities u_0, κ_0 are the velocity and diffusivity at the reference distance y_0 from the boundary. The extension to uniformly sloping beaches

$$h = h_0(y/y_0) ,$$ (4.2)

has been given by Kay (1987). If there is a continuous volume discharge rate q at y_0, then the solution of equation (3.2) is

$$c = \frac{q}{N\, h_0\, u_0 y_0} \left[\frac{y}{y_0}\right]^{s/2} R \exp\left[-\frac{R}{N^2}\left[\left[\frac{y}{y_0}\right]^N + 1\right]\right] I_{s/N}\left[\frac{2}{N}R\left[\frac{y}{y_0}\right]^{N/2}\right]$$ (4.3a)

with

$$N = 2 + r - s\,,\, R = u_0\, y_0^2/\kappa_0\, x\,.$$ (4.3b,c)

(Kay (1987, equation 5.15)), where $I_{s/n}$ is a modified Bessel function.
   For turbulent flow the appropriate power-law scalings are

$$r = \frac{1}{2}\,,\, s = \frac{3}{2}\,,\, N = 1$$ (4.4)

i.e. the velocity and turbulent diffusivity increase with increasing water depth. In this case Kay (1987) draws the important implications that as the outfall pipe is lengthened, the distance downstream at which the maximum shoreline concentration occurs increases as y_0, while the value of the maximum shoreline concentration decreases as $y_0^{-5/2}$. Thus a one-third increase in the offshore distance of a steady discharge

reduces the peak concentration at the shoreline by a factor of two. Similar results for tidal waters have been published by Macqueen and Preston (1983). Another important deduction made by Kay (1987) is that distributing the contaminant release along the length of the pipe is undesirable and increases the maximum shoreline concentration. So the best strategy is to discharge all the contaminant from the end of the pipe.

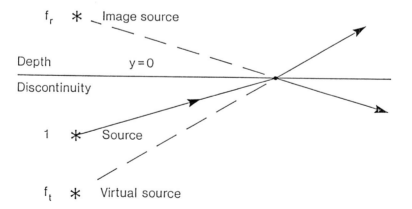

FIGURE 2. Sketch illustrating the positions and regions of contribution of the source, image source, and virtual source.

Another geometry which is amenable to exact solution is a depth discontinuity parallel to the flow direction. Kay (1985) shows that the solution can be obtained using the method of images (see Figure 2). In the lower region the solution comprises a source and image source contribution

$$c = \frac{q}{2[\pi u^{(-)} \kappa^{(-)} x]^{1/2}} [\exp(-\frac{-u^{(-)}(y + y_0)^2}{4\kappa^{(-)} x}) + f_r \exp(\frac{-u^{(-)}(y - y_0)^2}{4\kappa^{(-)} x})] \tag{4.5}$$

where f_r is the relative strength of the image (reflection). In the upper region the solution corresponds to there being a virtual source:

$$c = \frac{q}{2[\pi u^{(+)} \kappa^{(+)} x]^{1/2}} f_t \exp(\frac{-u^{(+)}(y + y_v)^2}{4\kappa^{(+)} x}), \tag{4.6}$$

where f_t is a transmission coefficient for contaminant across the depth discontinuity.
To match the x-dependence along the discontinuity, the virtual source needs to be positioned at

$$y_v = y_0 \left[\frac{\kappa^{(+)} u^{(-)}}{\kappa^{(-)} u^{(+)}}\right]^{1/2} \tag{4.7}$$

To match concentrations and fluxes across $y = 0$, Kay (1987, equation 3.7) calculates the relative strengths f_r, f_t of the image and the virtual sources to be

$$f_r = \frac{h^{(-)} (u^{(-)} \kappa^{(-)})^{1/2} - h^{(+)} (u^{(+)} \kappa^{(+)})^{1/2}}{h^{(-)} (u^{(-)} \kappa^{(-)})^{1/2} + h^{(+)} (u^{(+)} \kappa^{(+)})^{1/2}}, \tag{4.8a}$$

$$f_t = \frac{2 h^{(-)} (u^{(-)} \kappa^{(-)})^{1/2}}{h^{(-)} (u^{(-)} \kappa^{(-)})^{1/2} + h^{(+)} (u^{(+)} \kappa^{(+)})^{1/2}}, \tag{4.8b}$$

If the source is in the deeper water $(h^{(-)} > h^{(+)})$, then $f_t > 1$ and higher concentrations are found in the shallower water i.e. as is the case with the uniform beach solution (3.3) the plume bends towards the shallower water. Kay (1985) draws attention to the fact that the fluxes of contaminant show the opposite behaviour, and are greatest in the deep water. Hence, the higher concentrations in the shallower water are the consequence of weak transport of contaminant.

5 APPROXIMATE SOLUTIONS FOR STEADY DISCHARGES

Relative to horizontal length and velocity scales, the diffusivity κ_{22} in equation (3.3a) is extremely small (recall the narrowness of the plume). This disparity can be made explicit by formally introducing a small parameter ε:

$$m_1 \, m_2 \partial_t c + m_2 \, hu \, \partial_x c = \varepsilon \, \partial_y((m_1/m_2)h \, \kappa_{22} \, \partial_y c) . \tag{5.1}$$

For wave equations 'ray methods' exploit the presence of such a small parameter, leading to separate equations for the phase and amplitude of the waves. The rays are the principal directions, originating at the wave source, along which information propagates. Cohen and Lewis (1967) showed that similar ideas could be applied to diffusion problems. The specific case of discharge in rivers was addressed by Smith (1981b).

At small distances downstream the exact solutions (4.3a, 4.5) are exponential in character with ε-dependence:

$$c = A \, \exp(-\phi/\varepsilon) . \tag{5.2}$$

The ray approach enquires how the decay exponent ϕ and the amplitude factor A should vary if this were to be a global and not just a local form of solution.

Smith (1981b) shows how the behaviour of ϕ and A are related to the advection- diffusion vector K :

$$K = u/2\kappa_{22} . \tag{5.3}$$

If t is the local direction of a ray, which originated from the source, then along the ray

$$\frac{\partial \phi}{\partial s} = |K| - K \cdot t . \tag{5.4}$$

If n is the ray normal and κ the unit vertical vector, then the curvature ρ of the ray is given by

$$\rho = (n \cdot \nabla \, |K| + k \cdot (\nabla \times K))/|K| . \tag{5.5}$$

Thus, rays tend to go into regions of large $|K|$ and are bent in the sense of rotation of K (see Figures 3, 4) . The contaminant flux is in a direction halfway between t and K .

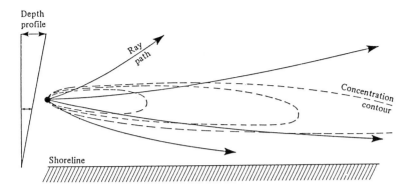

FIGURE 3. Ray paths for contaminant spreading when the water depth varies across the flow.

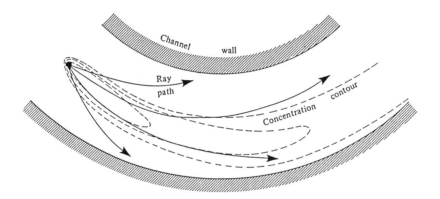

FIGURE 4. Ray paths for a source in a curved channel.

If J denotes the separation of adjacent rays, then the amplitude factor A has the solution

$$A J^{1/2} h^{1/2} |u|^{1/2} = \text{constant along rays.} \tag{5.6}$$

Thus, the contaminant concentrations tend to be greatest where the rays are close together, the water is shallow, or the current is weak.

6 INVERSE APPROACH

All the above work relates to the direct problem of calculating the consequences of a given discharge siting. Since men and their livestock inhabit the dry land, the most important part of the river, and of the concentration field, is the shoreline. For chimney plumes in the atmosphere the same considerations give predominance to ground-level concentrations. In the meteorological context F.B. Smith (1957) drew attention to a reciprocity relationship between an elevated source in the true wind, and a ground-level source in the reversed wind. In the first problem, the ground-level concentration is the same as the chimney-top concentration in the second problem. For alternative chimney heights the first problem would need to be solved repeatedly. However, for the inverse problem all that needs to be done is to evaluate the concentration at the alternative source positions.

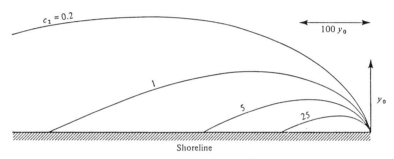

FIGURE 5. Kay's (1985) solution for the concentration experienced at the critical shoreline location as a function of the upstream discharge position. Note the 100:1 difference between the along- and cross-flow lengthscales.

Smith (1983) applies this same reciprocity principal to river discharges (see Figure 5). Despite the fact that the depth, diffusivity and velocity all go to zero at the shoreline, the inverse problem is still amenable to a ray solution. In particular, for separable geometry

$$h = h_0 \, f_1(x) \, f_2(y) \,, \, u = u_0 f_1(x)^{1/2} \, f_2(y)^{1/2} \tag{6.1}$$

the ray solution yields a maximum shoreline concentration which only depends upon the off-shore depth profile $f_2(y)$:

$$c_{max} = \frac{q}{h_0 u_0} \; \frac{10^{5/2} \, e^{-5/2}}{3 \pi^{1/2}} \; \left[f'_2(0)^2 \, f_2(y) \; \left[\int_0^y \, f_2(y')^{-1/2} \, dy' \right]^3 \right]^{-1} \tag{6.2}$$

This generalizes Kay's (1985) uniform beach result that the peak concentration at the shoreline is inversely proportional to the 5/2 power of the offshore distance y of the discharge.

The comparatively large power of the integral term in the denominator of (6.2) means that it is the full depth-topography that is important, and not just the depth at the discharge. In particular, a region of small depth (a sand bank) inshore of the discharge can profoundly reduce the shoreline concentration (see Figure 6).

7 STEADY DISCHARGES IN NARROW RIVERS

In the above work the river was implicitly assumed to be very wide, so all that mattered was the distance from the nearby shoreline. For narrow rivers the viewpoint is no longer appropriate and the river needs to be thought of as a whole.

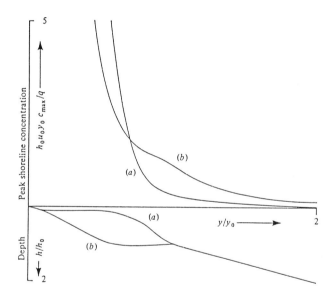

FIGURE 6. The peak concentration at the shoreline as a function of discharge distance away from the shoreline for depth profiles (a) with, and (b) without, offshore sandbanks.

As noted in the introduction, cross-sectional mixing takes place over a longitudinal length scale of several hundred times the channel breadth. For meandering rivers this can be more than the bend length. Thus, Yotsukura & Cobb (1972) laid great emphasis upon the importance of using flow-following coordinates. To measure the relative performance of different discharge positions across the flow they considered the degree of mixing:

$$P = 1 - 1/2 \ [|c - [c]|]/[c] \tag{7.1}$$

where the square brackets [...] indicate flux-weighted averages across the flow. Better discharges achieve nearly uniform concentrations more rapidly, and so have larger P values.

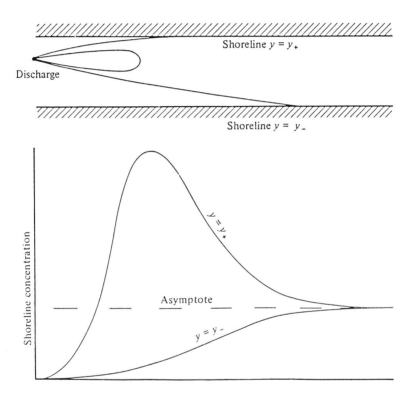

FIGURE 7.   Sketch of a contaminant plume in a river, and the concentration at the two shorelines.

Smith (1982) instead focussed attention upon the shoreline concentrations. If the discharge were too close to one of the banks the contaminant plume would reach the bank relatively soon with a high concentration (see Figure 7). For a point discharge at x_s, y_s the solution for c can be represented

$$c = q \ \sum_{n=0}^{\infty} \ \overset{u}{\phi}_n(x_s, y_s) \overset{d}{\phi}_n(x, y) \ \exp\left(-\int_{x_0}^{x} \mu_n(x') \ dx'\right) \tag{7.2}$$

where the upstream ϕ_n^u and downstream ϕ_n^d modes are defined

$$m_2 h u \partial_x \phi_n^d = \partial_y((m_1/m_2)\, h\kappa_{22}\partial_y\phi_n^d) + \mu_n m_2 h u\, \phi_{n'}^d$$

$$-m_2\, h u \partial_x \phi_n^u = \partial_y((m_1/m_2)h\kappa_{22}\, \partial_y\phi_n^u) + \mu_n m_2 h u \phi_n^u \, ,$$

with

$$h\kappa_{22}\partial_y\phi_n^d = h\kappa_{22}\, \partial_y\, \phi_n^u = 0 \;\; \text{on}\;\; y = y_-,\, y_+\, ,$$

$$\int_{y_-}^{y_+} m_2\, h u\, \phi_n^d\, \phi_n^u\, dy = Q = \int_{y_-}^{y_+} m_2\, h u\, dy\, ,$$

$$\int_{y_-}^{y_+} m_2\, h u\, \phi_n^u\, \partial_x\phi_n^d\, dy = \int_{y_-}^{y_+} m_2\, h u\, \phi_n^d \partial_x\, \phi_n^u\, dy = 0\, . \tag{7.3}$$

At large distances downstream the approach to uniformity is dominated by the slowest decaying mode $\phi_1^d(x, y)$. Smith (1982) infers that there is a concentration over-shoot at one of the banks unless this mode is eliminated. Hence the discharge position y_s across the flow must be at the single zero crossing of ϕ_1^u :

$$\phi_1^u(x_s, y_s) = 0\, . \tag{7.4}$$

To counteract the tendency for the plume to curve towards shallower water, the optimal site tends to be weighted towards the deeper water (see Figure 8). Similarly, to counteract the tendency for a plume to go to the outside of bends, the optimal site tends to be towards the inside at the beginning of a bend (see Figure 9).

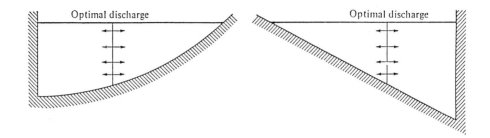

FIGURE 8. The positions of the optimal discharge sites in a semi-parabolic and in a triangular channel.

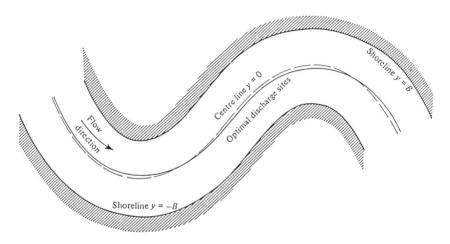

FIGURE 9. The optimal discharge site in a meandering river of parabolic cross-section.

8 SUDDEN DISCHARGES IN STRAIGHT RIVERS

For small-scale processes or accidential discharges, the contaminant is suddenly released into the river. For accumulative toxins it is the total exposure

$$a^{(0)} = \int_{-\infty}^{\infty} c \, dt \qquad\qquad (8.1)$$

that is important. Smith (1985) notes that formally $a^{(0)}$ satisfies the same equation as for steady discharges. Thus, the best site to avoid high concentrations at the shorelines is as given by equation (7.4).

More usually it is the peak concentration rather than the total exposure that is important. Close to the discharge high concentrations are unavoidable. At large distances (after cross-sectional mixing has been achieved) the dilution becomes diffusive in character Taylor (1953). However, as was noted by Aris (1956) the influence of the discharge position persists via a centroid displacement (which Aris did calculate) and a constant contribution to the variance of the concentration distribution (which Aris did not calculate).

For parallel flows exact results for the variance contribution from the source were first obtained by Gill and Sankarasubramanian (1971). Alas, their double infinite series solutions are rather cumbersome. Explicit results for the variance were derived by Chatwin (1977) for three special source distributions. The general result for arbitrary sources $q(y)$ was derived by Smith (1981a).

Contrary to the steady discharge situation, the variance at large distances downstream is maximized (and the concentrations minimized) if the discharge is made at the shallowest river bank. Physically the explanation is that the strong velocity shear encourages elongation and rapid dilution of the contaminant, while the low velocity gives more time before the pollution reaches other water users along the river (see Figure 10). The improvement in water quality by siting the discharge at the side rather than at the centre can be as much as a factor of two at distances large enough that mixing across the flow has been achieved.

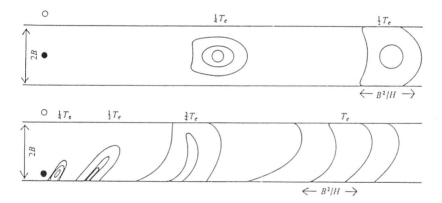

FIGURE 10. Sketch of contaminant clouds at equal time intervals for two different discharge positions across the flow. The e-folding time T_6 for mixing across the flow scales as B^2/HU_* and typically corresponds to a hundred channel breadths downstream.

Daish (1985) has investigated a variety of river depth profiles, and has shown that when there is a region of shallow water in the centre of a river it may be better to make sudden discharges there rather than at the side. Also, Daish (1985) has shown that further improvements in water quality far downstream can be achieved by splitting the discharge in two. The larger part of the discharge is at the bank where the channel bed slopes most gently, and the remainder near to where the channel is deepest. This is the optimal source distribution, so further subdivision is not necessary and, indeed, is undesirable. Peak concentrations are reduced by as much as a third as compared with the best single release site.

9 SUDDEN DISCHARGES IN VARYING RIVERS

As noted repeatedly throughout this review, the longitudinal length scale for mixing to have taken place can be comparable with or even greater than the bend length. Thus, the best location for a single discharge can jump from side to side of a river. More profoundly, Smith (1984) has shown that there exist optimal discharge sites which are better than all nearby sites upstream, downstream or across the flow. In keeping with the ideas protrayed in Figure 10 above, the optimal site is upstream of a region of shallow water.

By reciprocity there also exist optimal water intake sites which have minimum vulnerability to pollution from sudden discharges anywhere far upstream. These optimal intake sites are downstream of a region of shallow water.

10 CONCLUDING REMARKS

The main message of this paper is that it does make a substantial difference where a discharge or intake is sited. Also, the body of mathematical knowledge is now sufficiently great that comparisons between a range of alternative sites can be made, optimal sites can be precisely defined, and new discharge strategies can be suggested.

11 ACKNOWLEDGEMENTS

My attending of this symposium was made possible through the financial support of the Royal Society and of the Indian Institute of Technology, Kanpur.

12 REFERENCES

Aris, R., 1956. On the dispersion of a solute in a fluid flowing through a tube. Proc. R. Soc. Lond. A 235: 67-77.

Chatwin, P.C., 1977. The initial development of longitudinal dispersion in straight tubes. J. Fluid Mech. 80: 33-48.

Cohen, J.K. and Lewis, R.M., 1967. A ray method for the asymptotic solution of the diffusion equation. J. Inst. Math. applic. 3: 266-290.

Daish, N.C., 1985. Optimal discharge profiles for sudden releases in steady, uniform open channel flow. J. Fluid Mech. 154: 303-321.

Fischer, H.B., 1969. The effect of bends on dispersion in streams. Water Resouces Res. 5: 496-506.

_____, List, E.J., Koh, R.C.Y., Imberger, J. and Brooks, N.H., 1979. Mixing in Inland and Coastal waters. Academic Press.

Gill, W.N., and Sankarasubramanian, R., 1971. Dispersion of a non-uniform slug in time-dependent flow. Proc. R. Soc. Lond. A. 32: 101-117.

Kay, A., 1987. The effect of cross-stream depth variations upon contaminant dispersion in a vertically well-mixed current. Estuarine, Coastal & Shelf Science. (In the press) .Lauwerier, H.A., 1954. Diffusion from source in a skew velocity field. Appl. Sci. Res. 4: 153-156.

Macqueen, J.F. and Preston, R.W.. Cooling water discharges into a sea with a sloping bed. Water Res. 17: 389-395.

Smith, F.B., 1957. The diffusion of smoke from a continuous elevated point source into a turbulent atmosphere. J. Fluid Mech 2: 49-76.

Smith, R., 1979. Calculation of shear dispersion coefficients. In: Mathematical Modelling of Turbulent Diffusion in the environment. Academic Press, 343-363.

_____, 1981a. The importance of discharge siting upon contaminant dispersion in narrow rivers and estuaries. J. Fluid Mech. 108: 45-53.

_____, 1981b. Effect of non-uniform currents and depth variations upon steady discharges in shallow water. J. Fluid Mech. 110: 373-380.

_____, 1982. Where to put a steady discharge in a river. J. Fluid Mech. 115: 1-11.

_____, 1983. The dependence of shoreline contaminant levels upon the siting of an effluent outfall. J. Fluid Mech. 130: 153-164.

_____, 1984. Temporal moments at large distances downstream of contaminant releases in rivers.J. Fluid Mech. 140: 153-174.

_____, 1985. Contaminant dispersion as viewed from a fixed position. J. Fluid Mech. 152: 217-233.

Taylor, G.I., 1953. Dispersion of soluble matter in solvent flowing slowly through a tube. Proc. Roy. Soc. Lond. A. 219: 186-203.

Yotsukura, N., and Cobb, E.D., 1972. Transverse diffusion of solutes in natural streams. U.S. Geol. Survey 582-C.

Yotsukura, N. and Sayre, W.W., 1976. Transverse mixing in natural channels. Water Resources Res. 12: 695-704.

ANALYTICAL SOLUTION OF 3-D UNSTEADY STATE DIFFUSION EQUATION FOR A POLLUTANT FROM A POINT SOURCE DISCHARGE IN OFFSHORE REGION

Dr. V.P. SHUKLA
Computer Division, CWPRS, PUNE-411024, INDIA.

1 ABSTRACT

The analytical solution of the unsteady state three-dimensional advection-diffusion equation describing the diffusion of a non-conservative pollutant from a constant point source in two directions other than that in which tidal advection is predominant has been delineated. Model formulations include first order decay of the pollutant and reflective boundary conditions in the vertical direction.

The maximum length of pollution in the direction of the shore parallel current has been determined to be the ratio of twice the maximum advection velocity and frequency of the tidal oscillation. The maximum pollution level has been noted to occur at the time of one-fourth of the tidal period after which the pollution decreases rapidly. For a particular example, the vertical heterogeneity in the pollutant concentration lasts for only half of the tidal period. The lateral distribution of the pollutant is of Gaussian type with a rapid lateral decrease for small values of lateral diffusivity coefficient.

2 INTRODUCTION

The increasing industrial and municipal uses of the water bodies of this country are posing a dangerous problem of deterioration of the water quality, a hazardous condition for health and irrigational purposes. The harbour regions near the developed and developing cities are no exception to this problem. In these regions, the industrial and sewage outfalls are normally located at some distance offshore by means of long pipes (1-2 kms) placed perpendicular to the shore line. The dynamics of such a discharge pollutant may be considered as being released from a point source. This can appropriately be described by an unsteady three-dimensional advection-diffusion equation. Several numerical methods (Bredehoeft and Pinder (1973); Henry and Foree (1979); Holly and Preissman (1977); Konikow and Bredehoeft (1974); Pinder (1973)) have been developed to solve such equations. An advantage of such numerical approaches is that heterogeneity of the medium, complicated boundary conditions, and irregular domains can also be taken into account without much difficulty. However, these models require a plethora of data and numerical costs can be very expensive. Also, the lack of sufficient initial data often renders the use of these models inappropriate. In such cases, analytical solutions can be used to obtain the necessary approximate initial data. They can also be used to develop the numerical methods and to decide the suitability of a numerical procedure for a given problem.

The analytical solutions for three-dimensional diffusion-advection problems in water bodies have been discussed by Kuo (1976), Yeh and Tsai (1976) and Prakash (1977) for point source discharges of heat or some other pollutant. Recently, a three-dimensional dispersion problem with vertical leaching from an area source has employed Green's function method (Sagar (1982)) to obtain approximate analytical solutions.

It may be noted that the aforesaid analytical models assume the convective velocity of the pollutant to be unidirectional and uniform, and thus are applicable to the pollution problems in streams, rivers and lakes only. For the models to be applicable for estuaries or harbour regions the convective velocity should be considered to be time-dependent (say almost periodic). In this paper, we consider the three-dimensional unsteady state diffusion of the pollutant released from a point source and convected by a shore parallel current (convective velocity being periodic with respect to time). The solution is obtained by using Fourier transform method. Concentration profiles with respect to both time and space variables are discussed.

3 THE POLLUTANT DISPERSION EQUATION

Let $c(x, y, z, t)$ be the concentration of a pollutant driven along x-axis under tidal influence. Assume the pollutant satisfies a first order radioactive decay. The dynamics of the pollutant can then be described by the following advection-diffusion equation:

$$\frac{\partial c}{\partial t} + u \frac{\partial c}{\partial x} = \frac{\partial}{\partial x}(D_L \frac{\partial c}{\partial x}) + \frac{\partial}{\partial y}(D_H \frac{\partial c}{\partial y}) + \frac{\partial}{\partial z}(D_V \frac{\partial c}{\partial z}) - kc \tag{1}$$

where $u = u_0 \sin \sigma t$, u_0 being the maximum tidal velocity; σ, the frequency of oscillation; k, the constant rate of decay of the pollutant; D_L, D_H and D_V, the longitudinal, lateral and vertical coefficients of diffusion, assumed to be constants for the sake of simplicity. Since the tidal velocity u is dominant in the x-direction, the effect of diffusion in the x-direction may be regarded as negligible.

We consider the source to be a point located at $x = 0$, $y = y_0$, $z = z_s$. This can be modelled by the condition

$$c(0, y, z, t) = M \, \delta(y - y_0) \, \delta(z - z_s) \tag{2}$$

where $\delta(\bullet)$ is the Dirac delta function and M, the quantity of effluent injected per unit time at $y = y_0$ and $z = z_s$, is given by

$$M = C_0 Q_0 / u_0 ,$$

where C_0 is the concentration of the pollutant in the discharged effluent and Q_0 is the volume of the effluent injected per unit time (Prakash (1977)). This boundary condition corresponds to the offshore discharge of sewage and other wastes from a submerged pipe line, placed perpendicular to the shore line, at the depth z_s from the ocean surface.

At infinite lateral distance the concentration tends to zero,

$$c(x, y, z, t) \to 0 \text{ as } y \to \pm\infty \;.$$

(3)

This condition requires that when the point source is located at a considerable distance from the shore then the pollution is the least in the near shore region.

Also, we consider the following reflective boundary conditions

$$-D_V \, \partial c / \partial z = 0 \text{ at } z = 0 \;.$$

(4)

$$-D_V \, \partial c / \partial z = 0 \text{ at } z = h \;.$$

(5)

The former condition means that the pollutant cannot escape the water surface, while the latter signifies the existence of the horizontal base at depth $z = h$, say, beyond which the vertical mixing is not possible.

In addition, we prescribe the following initial condition

$$c(x, y, z, 0) = f(x, y, z)$$

(6)

To cast the problem in dimensionless form, we define the dimensionless time and space variables as

$$T = \frac{D_V t}{h^2}, \; X = \frac{D_V x}{u_0 h^2}, \; Y = \frac{y}{h}, \; Z = \frac{z}{h} \;.$$

(7)

Also, we define the dimensionless concentration as

$$C = \frac{h^2 c}{M}$$

(8)

Then, equations (1)-(6) can be written in dimensionless form as

$$\frac{\partial C}{\partial T} + \sin \mu T \frac{\partial C}{\partial X} = \alpha \frac{\partial^2 C}{\partial X^2} + \beta \frac{\partial^2 C}{\partial Y^2} + \frac{\partial^2 C}{\partial Z^2} - \gamma C$$

(9)

$$C(0, Y, Z, T) = \delta(Y - Y_0)\, \delta(Z - Z_s) \tag{10}$$

$$C(X, Y, Z, T) = 0 \text{ as } Y \to \pm \infty \tag{11}$$

$$-\partial C/\partial Z = 0 \qquad \text{at } Z = 0 \tag{12}$$

$$-\partial C/\partial Z = 0 \qquad \text{at } Z = 1 \tag{13}$$

$$C(X, Y, Z, 0) = \frac{2h}{M}\, f(X, Y, Z) \ . \tag{14}$$

where

$$\alpha = \frac{D_L D_V}{u_0^2 h^2}, \ \beta = \frac{D_H}{D_V}, \ \gamma = \frac{h^2 k}{D_V}, \ \delta = \frac{h^2 \sigma}{D_V}, \tag{15}$$

It can be seen from the expression for α given in (15) that since D_L, D_V are very small in comparison to the value of u_0, α may be regarded as negligible, and therefore (9) is approximately

$$\frac{\partial C}{\partial T} + \sin \mu T \ \frac{\partial C}{\partial X} = \beta \frac{\partial^2 C}{\partial Y^2} + \frac{\partial^2 C}{\partial Z^2} - \gamma C \tag{16}$$

To solve equation (16) under conditions (10)-(14), we proceed as follows.

4 ANALYTICAL SOLUTION

Applying Fourier's transforms (exponential type with respect to Y and cosine type with respct to Z) on (16), (10) and (14) under boundary conditions (11)-(13), we have

$$\frac{\partial \overset{\triangle}{C}}{\partial T} + \sin \mu T \ \frac{\partial \overset{\triangle}{C}}{\partial X} = -b \overset{\triangle}{C} \tag{17}$$

$$\overset{\triangle}{C}(0, \omega, n, T) = e^{i\omega Y_0} \cos(n\pi Z_s) \tag{18}$$

$$\overset{\triangle}{C}(X, \omega, n, 0) = \frac{h^2}{M} \overset{\triangle}{f}(X, \omega, n) \tag{19}$$

where

$$b = \beta \omega^2 + n^2 \pi^2 + \gamma \qquad (20)$$

$$\overset{\triangle}{C}(X, \omega, n, T) = \int_0^1 [\int_{-\infty}^{\infty} C(X, Y, Z, T) \, e^{i\omega Y} \, dY] \cos(n\pi Z) dZ \qquad (21)$$

ω and n being infinite and finite Fourier transform parameters.
Using the transformation

$$\overset{\triangle}{C}(X, w, n, T) = S(X, w, n, T) \, \overline{e}^{bt} \qquad (22)$$

on (17)-(19), we get

$$\frac{\partial S}{\partial T} + \sin \mu T \frac{\partial S}{\partial X} = 0 \qquad (23)$$

$$S(0, \omega, n, T) = e^{bT + i\omega Y_0} \cos(n\pi Z_s) \qquad (24)$$

$$S(X, w, n, 0) = \frac{h^2}{M} \overset{\triangle}{f}(X, \omega, n) \qquad (25)$$

Also, on substituting

$$T = \mu^{-1} \cos^{-1}(1 - \mu\tau) \qquad (26)$$

in (23)-(25), we obtain

$$\frac{\partial S}{\partial \tau} + \frac{\partial S}{\partial X} = 0 \; ,$$

$$S(0, \omega, n, \tau) = \exp[b\mu^{-1} \cos^{-1}(1 - \mu\tau) + i\omega Y_0] \cos(n\pi Z_s)$$

$$S(X, \omega, n, 0) = \frac{h^2}{M} \overset{\triangle}{f}(X, \omega, n)$$

which can be solved to yield

$$S(X, \omega, n, \tau) = \frac{h^2}{M}\, \hat{f}(X - \tau, \omega, n) \quad \text{for } \tau < X$$

$$S(X, \omega, n, \tau) = \exp[b\mu^{-1} \cos^{-1}(1 - \mu(\tau - X)) + i\omega Y_0]\cos(n\pi Z_s) \quad \text{for } t \geq X.$$

Using (26), these can be written in terms of T and X as

$$S(X, \omega, n, T) = \frac{h^2}{M}\, \hat{f}\left(X - \frac{1 - \cos\mu T}{\mu}, \omega, n\right) \text{ for } 1 - \cos\mu T < \mu X \tag{27}$$

$$S(X, \omega, n, T) = \exp\left[\frac{b}{\mu}\cos^{-1}(\mu X + \cos\mu T) + i\omega Y_0\right]\cos(n\pi Z_s) \text{ for } 1 - \cos\mu T \geq \mu X \tag{28}$$

Substituting (27)-(28) in (22) and applying inverse Fourier exponential transform with respect to ω, we get

$$\hat{C}(X, Y, n, T) = \frac{h^2 e^{-(n^2\pi^2 + \gamma)}}{2M(\pi\beta T)^{1/2}} \int_{-\infty}^{\infty} e^{-(y-\xi)^2 (4\beta T)}\, \hat{f}\left(X - \frac{1 - \cos\mu T}{\mu}, \xi, n\right)d\xi$$

$$\text{for } 1 - \cos\mu T < \mu X \tag{29}$$

$$\hat{C}(X, Y, n, T) = \frac{e^{\{(n^2\pi^2 + \gamma)(T - \phi) + (Y - Y_0)^2/[4\beta(T - \phi)]\}}}{2M(\pi\beta T)^{1/2}}\cos(n\pi Z_s)$$

$$\text{for } 1 - \cos\mu T \geq \mu X \tag{30}$$

where

$$\phi = \mu^{-1}\cos^{-1}(\mu X + \cos\mu T).$$

Now applying the inverse Fourier cosine transform with respect to n, we get the solution $C(X, Y, Z, T)$ as

$$C(X, Y, Z, T) = \hat{C}(X, Y, 0, T) + 2\sum_{n=1}^{\infty} \hat{C}(X, Y, n, T)\cos(n\pi Z) \tag{31}$$

where $\hat{C}(X, Y, Z, T)$ is as given by (29)-(30).

Using the dimensionless solution $C(X, Y, Z, T)$ given by (31), and equations

(7)-(8), the dimensional solution $c(x, y, z, t)$ can be given as

$$c(x, y, z, t) = \hat{c}(x, y, 0, t) + 2 \sum_{n=1}^{\infty} \hat{c}(x, y, n, t) \cos(n\pi z/h) \qquad (32)$$

where

$$\hat{c}(x, y, z, t) = \frac{h}{2(\pi D_H t)^{1/2}} \exp[-(k + \frac{n^2 \pi^2 D}{h^2} v)t] \times$$

$$\times \int_{-\infty}^{\infty} \exp[-\frac{(y - \xi)^2}{4 D_H t}] \hat{f}(\frac{D v x}{u_0 h^2} - \frac{1 - \cos \mu T}{\mu}, \xi, n) d\xi$$

$$\text{for } 1 - \cos \sigma t < \frac{\sigma}{u_0} x \qquad (33)$$

$$\hat{c}(x, y, z, t) = \frac{M}{2h(\pi D_H (t - \theta))^{1/2}} \exp[-(k + \frac{n^2 \pi^2 D}{h^2} v)(t - \theta)] \times$$

$$\times \exp[-\frac{(y - y_0)}{4 D_H (t - \theta)}] \cos(n\pi z_s/h)$$

$$\text{for } 1 - \cos \sigma t \geq \frac{s}{u_0} x \qquad (34)$$

where

$$\theta = s^{-1} \cos^{-1}(\frac{\partial x}{u_0} + \cos \sigma t) \qquad (35)$$

In the following, we shall discuss the nature of the solution given by (32)-(34) with respect to time and space variables for a particular example.

5 AN APPLICATION

Consider a pollutant which is being continuously discharged at a point on z-axis , 1 m below the water surface into an oscillating flow with a period of 12.4 hours. Thus, the frequency of oscillation is $\sigma = 2\pi/44600$ rad./sec. = 0.5071629 rad./hr. The concentration of the pollutant in the effluent, C_0 , is 2658.69 ppm , and the volume of effluent released per unit time, Q_0 , is $0.23 \text{ m}^3/\text{sec}$ (828 m^3/hr) . The initial concentration of the pollutant in the water body is assumed to be zero and other parameters are

$$D_H = 22 \text{ m}^2/\text{sec} = 79200 \text{ m}^2/\text{hr}$$

$$D_V = .076 \text{ m}^2/\text{sec} = 273.6 \text{ m}^2/\text{hr}$$

$$u_0 = .275 \text{ m/sec} = 990 \text{ m/hr}$$

$$h = 22\text{m}, \quad k = .005 \text{ per hr.}$$

Using the above data (Prakash (1977)), the solution given by (32)-(34) has been plotted as a function of $t, x,$ and z . Figure 1 shows the variation of the pollutant concentration with time at two different locations $(100, 0, 1)$ and $(200, 0, 1)$. The pollution level increases at first and at about the third hour (almost one-fourth of the tidal period) reaches the maximum value after which it starts to decrease, probably due to a decrease in the tidal velocity. We note that the peak concentration at location $(100, 0, 1)$ is higher than that observed at location $(200, 0, 1)$, and that the descends on both sides of the peak concentration at the former location is steeper than the descends at the latter location. This shows that increase in the pollution level near the region of the point source is rapid and the decrease after the peak concentration is also rapid. It is evident from Figure 1 that by the eleventh hour the pollution is considerably reduced but does not vanish. It may be noted that zero value of the pollution at 12 and 13 hours is obtained due to our assumption of zero initial concentration in the water body (i.e. $f(x, y, z) = 0$ in (6) due to which c is given by (33) and thus c as given by (32) is zero) Again, as the time advances, some concentration of the pollutant is noticed (Figure 1) but it continues to decrease very slowly.

As mentioned earlier in this example, the solution is zero for $1 - \cos \sigma t < (\sigma/u_0)x$, and therefore behaviour of the solution given by (32) with space variables depends only on c given by (34) for $1 - \cos \sigma t \geq (\sigma/u_0) x$. The latter condition yields the maximum value of x , say x_{max} , where the pollution is expected to be nonzero, as $2 u_0/\sigma$ and this is obtained at $t = \pi/\sigma$ i.e. half of the tidal period. It can be seen from Figure 2 that pollutant concentration decreases along the shore parallel distance x and traverses the distance 900, 1800 and 3800 meters in 2, 3, and 6 hours respectively. The maximum distance travelled by the pollutant in this example is $(x_{max} = 2u_0/\sigma)$ 3904.07 m , and the time elapsed is 6.2 hr after which the pollution is expected to retreat due to the current reversal. This is apparent from Figure 2 in which the curve corresponding to the time 8 hours indicates the total distance travelled as 3100 m only .

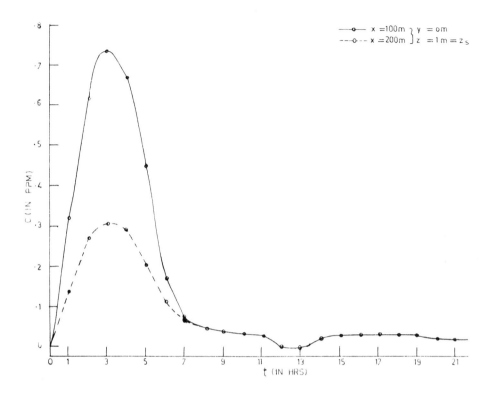

FIGURE 1 Concentration profiles of the Pollutant as a function of time at y = 0m , z = 1n
for x = 100m and 200m .

The concentration profiles of the pollutant as a function of vertical distance have
been shown in Figure 3 for the time 2, 3, 6 and 8 hours. The curves for 2 and 3 hours
signify the decrease of the concentration of the pollutant with the increase in the vertical
distance. Both the curves are changing from concave down to concave up, and there
exists a point of inflection on either curve. As time increases, the decrease in the
pollutant concentration with the vertical distance becomes less pronounced and after
some time the concentration may remain constant for all z (see Figure 3, the curves for
t = 6 and 8 hours). This is due to large values of vertical diffusivity coefficient, as is
noted from (34).

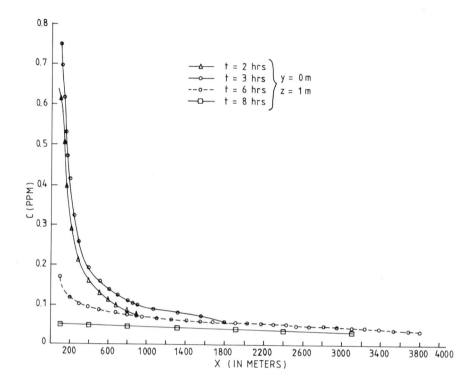

FIGURE 2 Concentration Profiles of the Pollutant as a Function of the Shore Parallel
Distance x at y = 0m , z = 1m for t = 2, 3, 6 and 8 hours

As is evident from the solution (32)-(34), the y-dependence of the pollutant concentration is Gaussian and thus symmetric with respect to y_0. The smaller the diffusivity constant D_H, the higher will be the concentration peak at $y = y_0$ and the steeper will be the descends on both sides. This implies that the smaller the lateral diffusion, the smaller will be the concentration of the pollutant near the shore region. In the example under consideration, the pollutant concentration at points (100, 500, 1) , (100, 1000, 1) , (100, 1500, 1) and (100, 2000, 1) have been computed to be .0298, .0235, .0157 and .0089 ppm at t = 11 hours. Depending on the permissible quantity of a particular pollutant such values can be useful in determining the proper location of the outfall so as to produce the least detrimental effects in the near shore regions.

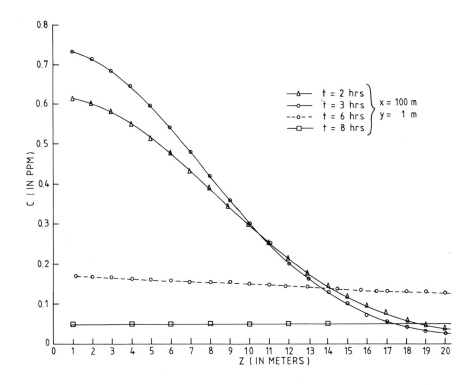

FIGURE 3 Concentration Profiles of the Pollutant as a Function of Vertical Distance z
at x = 100m , y = 0m for t = 2, 3, 6, and 8 hours

6 ACKNOWLEDGEMENTS

The author is thankful to Mr. M.T. Limaye (LAI) , Mr. N.N. Tambe (Tracer) and Mr.
N. Prasad (ARO) for drawing the concentration profiles and discussing the nature of
these profiles.

7 REFERENCES

Bredehoeft, J.D., and Pinder, G.F., 1973. Mass transport in flowing ground water.
    Water Resources Research, Vol. 9, No. 1, pp. 194-210.
Henry, J.F., and Foree, E.G., 1979. Dispersion modeling in time and two-
    dimensions. Journal of Environmental Engineering Division, ASCE, Vol.
    105, No. EE6, Proc. Paper 15054, pp. 1131-1147.
Holly, Jr., F.M., and Preissmann, A., 77. Accurate calculation of transport in
    two-dimensions. Journal of Hydraulic Division, ASCE, Vol. 103, No. HY 11,
    Proc. Paper 13336, pp. 1259-1277.
Koniknow, L.F., and Bredehoeft, J.D., 1974. Modeling flow and chemical quality
    changes in an irrigated stream aquifer systems. Water Resources Research,
    Vol. 10, No. 3, pp. 546-562.

122

Kuo, E.Y.T., 1976. Analytical solution for 3-D diffusion model. Journal of the Environmental Engineering division, ASCE, Vol. 102, No. EE4, Proc. Paper 12315, pp. 805-820.

Pinder, G.F., 1973. A Galerkin finite element simulation of ground water contamination of Long island, New York. Water Resources Research, Vol. 9, No. 6, pp. 1657-1669.

Prakash, A., 1977. Convective-dispersion in perennial streams. Journal of Environmental Engineering Division, ASCE, Vol. 3, No. EE2, Proc. Paper 12891, pp. 321-340.

Sagar, B., 1982. Dispersion in three-dimensions: Approximate analytic solutions. Journal of Hydraulic division, ASCE, vol. 108, No. HY1, Proc. Paper 16806, pp. 47-62.

Yeh, G. and Tsai, Y., 1976. Analytical three-dimensional modeling of efflulent discharges. Water Resources Research, Vol. 12, No. 3, pp. 533-540.

PART III - POPULATION ECOLOGY

MODELLING SURVIVAL IN CHEMICALLY STRESSED POPULATIONS

THOMAS G. HALLAM
Department of Mathematics and Program in Ecology, University of Tennessee,
Knoxville, Tennessee U.S.A.

1 ABSTRACT

Any study of effects of a toxicant on a population should include an assessment of survival of the population. Survival analyses in several toxicant-population models are surveyed here to indicate different parameters that may be employed to estimate persistence or extinction of the stressed populations. A comparison of analytical results for extinction in both continuous and discrete population models are addressed from a modelling perspective. Roles of fluctuating demographic parameters and density dependence in survival are emphasized. The models generally contain state variables representing the population biomass, the concentration of toxicant in the environment, and the concentration of toxicant in the organism. The continuous results are indicated for a Smith-like population model while the discrete model results are for the logistic equation.

2 INTRODUCTION

The utility, indeed the necessity, of mathematical modelling in assessing the effects of chemicals on a biological population is readily demonstrated by a study of almost any set of empirical data. Because of incompleteness and difficulties in making long term studies, accurate extrapolation is usually not feasible. Chemical fate models are the basis of most current risk assessment schemes with the EXAMS (Lassiter et. al. (1978)) model being perhaps the most widely used in the U.S. A comparison of the fate models EXAMS, PEST (Park et. al. (1980)) and FOAM (Bartell et. al. (1984)) may be found in a report of a U.S. Environmental Protection Agency workshop in ecotoxicology (Levin et. al. (1984)). Other recent articles which discuss state of the art modelling efforts in ecotoxicology include the 1981 EPRI Workshop on Cycling and Effects of Toxic Substances (EPRI EA,1988), Ecotoxological Test Systems (Hammons (1981)) and a special issue of Ecological Modelling, Vol. 22, 1984.

An assessment of effects of a hazardous chemical on a population must be cognizant of the environment and other associated stresses, the genetic composition, the size-stage structure, and the spatial distribution of the population. Obviously, current assessment tools do not completely account for these and other important factors. However, in my opinion, a determination of risk should minimally include a determination of population survival.

The methodology and approach to most population-toxicant interactions has been to utilize simulation techniques. A purpose of this article is to take a different approach and to integrate some recent research on population models where analytical

126

techniques have been utilized to derive generic persistence-extinction criteria. These criteria are based upon population level parameters that might be useful in hazard assessment. I will review some related literature and discuss applicablility modes. As chemical models are reasonably well developed, I will primarily concentrate on the suitability of a population model for estimation of survival.

3 GENERIC MODELS OF POPULATION-TOXICANT INTERACTIONS

Rudimentary, but generic, population models coupled with first order chemical kinetic equations have been recently employed to study population-toxicant interactions (Hallam et. al. (1983, 1984)). The solution techniques employed in these articles are fruitful because the analytical derivations propose general principles that have uniform application. An indication of the type of conclusions and its utility for hazard assessment is now presented. The models considered here are usually composed of three differential equations: one representing the population dynamics, one representing the dynamics of the concentration of toxicant in the organism and one representing the dynamics of the concentration of toxicant in the environment. These variables are coupled by a dose-response formulation.

To indicate the character of the results, the following variables are specified: $x = x(t)$, the biomass of the population at time t; $c_0 = c_0(t)$, the concentration of toxicant in the organism at time t; and $c_E = c_E(t)$, the concentration of toxicant in the environment at time t. The dynamics of the population are assumed given by

$$\frac{dx}{dt} = \frac{x}{g(x)} [r(c_0) - f(x)x], \ x(0) > 0,$$ (1)

where f is a continuous function from $R_+ = [0, \infty)$ into R_+ and g is a continuous function defined on R_+ and whose range is a subset of R_+ that does not contain zero. Sufficient smoothness on g and f to guarantee uniqueness to solutions of initial value problems is assumed and it is tacitly assumed that solutions of (1) are bounded on R_+ This formulation contains rudimentary equations such as the logistic and Smith's model (1963). The organismal toxicant dynamics are assumed given by first order chemical kinetics:

$$\frac{dc_0}{dt} = k_1 c_E - g_1 c_0 - mc_0 + h(t), \qquad c_0(0) = 0.$$ (2)

The first two terms on the right side of (2) represent net uptake of toxicant from the environment, the third term represents depuration, and $h = h(t)$, represents uptake of toxicant through the food chain pathway.

The environmental toxicant dynamics can be derived by employing a toxicant balance equation (Hallam et. al. (1983)) to yield

$$\frac{dc_E}{dt} = -k_2 c_E x + g_2 c_0 x - n_2 c_E + e(t), \qquad c_E(0) = 0. \qquad (3)$$

The first two terms on the right are consequences of the first order chemical kinetics hypothesis and represent removal of the toxicant present in the environment by the population, the third term represents losses of chemical from the environment to exogeneous components, and the final term is input concentration of toxicant into the environment. The demographic dose-response coupling $r(c_0)$ is usually taken as linear: $r(c_0) = r_0 - r_1 c_0$. This hypothesis is a "worst case" situation (de Luna (1983)) from the perspective that knowledge of the behavior of other system with a linear dose-response function provides a lower bound on population levels for systems with other classical type dose-response formulations such as hyperbolic or sigmoid types (e.g. Butler (1978); Filov (1979)).

Results for model {(1), (2), and (3)} have focused on persistence and extinction phenomena for the biological population. Survival will depend upon the degree of perturbation that organismal toxicant concentration causes in the demographic parameter r.

Some definitions (e.g. Freedman and Waltman (1984, 1985); Hallam and Ma (1987)) that help delineate the asymptotic behaviors of the population are now given. A solution, $x = x(t)$, of (1) is called a population. A population, x, is weakly persistent provided $\limsup_{t \to \infty} x(t) > 0$; it is persistent if $\liminf_{t \to \infty} x(t) > 0$; and it is persistent in mean provided $\liminf_{t \to \infty} t^{-1} \int_0^t x(s) \, ds > 0$. The population goes to extinction if it is not weakly persistent. Because of the assumed uniqueness of initial value problems, no finite time extinction can occur in (1).

To contrast results of this section with some of the work discussed subsequently on populations with fluctuating demographic parameters, asymptotic persistence-extinction criteria are indicated.

The equation (1) is restricted, for convenience of illustration, to be a modified Smith's equation (Hallam and de Luna (1984))

$$\frac{dx}{dt} = \frac{x[r(c_0) B(r_0 - c - a) - acx]}{B(c + r_0 - a) + ax}, \qquad x(0) > 0, \qquad (1a)$$

where B, c, and a are constants and, as above, $r(c_0) = r_0 - r_1 c_0$. In the following, for an arbitrary function $f(t)$, f^* will designate the upper limit, $\limsup_{t \to \infty} f(t) = f^*$, and f_* will designate the lower limit, $\liminf_{t \to \infty} f(t) = f_*$. The solutions of (1a) are bounded by $X_m = \max(x(0), K)$ where $K = r_0 B(r_0 - c - a)/ac$.

<u>Theorem 1</u>. Define $\rho^M = \dfrac{h^*}{m} + \dfrac{k_1}{n_2 m} (e^* + g_2 X_m)$ and

$$\rho_M = \dfrac{h^*}{m} + \dfrac{k_1 e^*}{m(k_2 X_m + n_2)}$$

(i) If $r_0 > r_1 \rho^M$ then the population modelled by (1a) is weakly persistent under the stress of toxicant as modelled by (2) and (3).

(ii) If $r_0 < r_1 \rho_M$ then the population modelled by (1a) goes to extinction under the stress of toxicant as modelled by (2) and (3).

This result yields persistence-extinction criteria that are functions of upper and lower bounds on organismal toxicant concentration and the <u>a priori</u> upper bound on the solutions. The result itself is new and can be obtained by a trivial extension of the techniques developed in Hallam and de Luna (1984) where uniform bounds on R_+ are assumed, rather than the asymptotic bound mentioned here. A disadvantage of this result is the utilization of the <u>a priori</u> bound in the system level survival parameters but with the nonlinear complexity of this population-toxicant model, this approach seems useful. As indicated below these results can be refined for oscillatory toxicant concentrations in that the <u>a priori</u> bound is not required; but before proceeding to fluctuating demographic parameters, we simplify the chemical dynamics.

<u>3.1 A Simplified Chemical Model</u>.

When the environmental medium is sufficiently large so that the environmental toxicant concentration remains virtually unchanged despite continued uptake and egestion by organisms, it is reasonable to assume that $k_2 = g_2 = 0$ and replace (3) by the model

$$\dfrac{dc_E}{dt} = e(t) - n_2 c_E , \qquad (3a)$$

(see Hallam <u>et. al.</u> (1983)). Numerical studies comparing {(1a), (2), (3)} and {(1a), (2), (3a)} support the application of this "reduced" model in that for risk assessment purposes toxicant concentrations and population estimates are on the "safe" side (Figures 1, 2) for estimating risk.

The reduced model where (3a) replaces (3) is also more tractable analytically than the original model, indeed, the problem is reduced to consideration of a single equation, the population model with time varying coefficients. For convenience of illustration the subsequent developments will all relate to the simpler formulation.

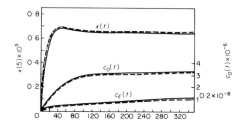

FIGURE 1 Behavior of the components of models (C: 1a, 2, 3) and (R: 1a, 2, 3a) .
Broken lines represent the components of model (C) . (From Hallam and de
Luna, 1984).

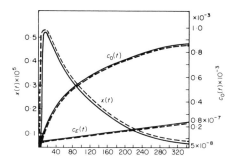

FIGURE 2 Extinction behavior of models (C) and (R) . Dashed curves represent the
components of model (C). (From Hallam and de Luna, 1984).

4 FLUCTUATIONS IN DEMOGRAPHIC PARAMETERS

Due to limitations of experimental techniques, ecological data are often described
by temporal averages. This policy is carried a step further here in that deterministic
system behavior is discussed in terms of component function means. This approach
leads to hypotheses that are stated in terms of the means of the time varying inputs.
Persistent fluctuations in functions can be measured asymptotically by successively
iterated means. Notation employed here for the nth-iterated mean of an oscillatory
function $p = p(t)$ is

$$\langle p^n \rangle (t; t_0) = (t - t_0)^{-1} \int_{t_0}^{t} \langle p^{n-1} \rangle (s; t_0)ds , \quad n = 1,2,...$$

where $\langle p^0 \rangle$ is the function $p(t)$ itself.

A recent result (Hallam and Ma (1986)) describes survival conditions for a
toxicant stressed population in terms of the first mean. The situation is applicable when
$c_0(t)$ does not have a limit; that is, when $c_0^* \neq c_{0^*}$.

Theorem 2. The population-toxicant system consisting of equations {(1), (2), and (3a)} with a linear dose response has the asymptotic behavior as follows:

(i) Let $c_0^U \equiv \dfrac{\langle h \rangle}{(g_1 + m)} + \dfrac{k_1}{n_2(g_1 + m)} \langle e \rangle*$ and $r_1 c_0^U < r_0$; then, the population is

weakly persistent and persistent in mean.

(ii) Let $c_0^L \equiv \dfrac{\langle h \rangle*}{(g_1 + m)} + \dfrac{k_1 \langle e \rangle*}{n_2(g_1 + m)}$ and $r_1 c_0^L > r_0$; then the population goes to

extinction.

It is also possible to obtain persistence results when all iterated means are oscillatory. The limits, $h^* = \lim_{n \to \infty} \langle h^n \rangle*$ and $e^* = \lim_{n \to \infty} \langle e^n \rangle*$ can be shown to exist (Hallam and Ma (1986)). In terms of these limits, the following persistence results hold.

Theorem 3. The reduced model (1), (2), and (3a) represents a persistent population provided

$$\frac{h^*}{(g_1 + m)} + \frac{k_1 e^*}{n_2(g_1 + m)} < \frac{r_0}{r_1}$$

The reader is referred to Hallam and Ma (1986) for the proofs of Theorems 2 and 3.

4.1 Summary on Fluctuations in Population-Toxicant Systems.

Fluctuations in the demographic parameters of the above population-toxicant model are reflections of temporal oscillations in the exogeneous toxicant inputs, modelled by the functions h and e and may or may not lead to oscillations in the population behavior. In any case, the above results yield a hierarchy of persistence criteria that will apply to a wide class of population models although for illustration we have restricted our considerations to a single model. If the organismal toxicant concentration, c_0 , approaches a limit, $\lim_{t \to \infty} c_0(t) = c_\infty$, then $c_\infty = r_0/r_1$ is the threshold value that separates persistent models where all populations survive from models in which all the populations go to extinction. If $\lim_{t \to \infty} c_0(t)$ does not exist but

the limit, $\lim_{t \to \infty} \langle c_0^1 \rangle (t, 0) = \langle c_0^1 \rangle_\infty$ does exist (as occurs for example, if c_0 is a periodic function of t),then $\langle c_0^1 \rangle_\infty = r_0/r_1$ is the extinction threshold separating persistence models from those where extinction occurs. The hierarchy may be continued if oscillations reoccur in higher order means.

The existence of the extinction threshold and knowledge of its value, important information in developing formulations for the effects of hazardous chemicals on populations, can provide guidance for assessing risks associated with exposure. To impliment these results, only the intrinsic growth rate in the absence of the toxicant, r_0 , and the toxicant concentration response parameter, r_1 , need to be established. A major contribution of these studies is that demographics are fundamental to determining persistence or extinction of the population while environmental parameters are not as fundamental in that they determine levels of persistence.

5 DENSITY DEPENDENT SURVIVAL ANALYSES

The above survival analyses while accurate mathematically are possibly misleading from an assessment perspective. A main conclusion is that persistence can be determined solely by assessing the role of the organismal toxicant concentration in perturbing the demographic parameters. The results also indicate that density dependence plays no role in assessing survival. Due to the definition of persistence, at zero density, density dependent effects are not important. Extinction, as previously defined, means model trajectories approach zero; however, if, for example, the population, x , is measured in biomass, then for a population to be viable, x must always exceed the neonatal biomass at which an individual has a positive probability of survival. There are other indicators, determined possibly by both biological and economical constraints, that might be called, or result in, extinction when the population level is not zero (Hallam and Ma (1987)). In such cases, it is expected that if the model is defined on an infinite time horizon, that even these indicators of extinction ultimately mean model trajectories approach zero.

These ideas lead naturally to implimenting the concept of finite time extinction. Hallam and Ma (1987) have argued that for a generic, smooth, population model such as (1), if solutions to zero inital value problems are unique, there is no possibility of an extinction, in the sense that a population approaches zero, on a finite time horizon. In particular, to have finite time extinction, one must violate uniqueness of zero initial value

extinction, one must violate uniqueness of zero initial value problems. A natural hypothesis $(\int_0 \frac{g(x)}{x} \, dx = \infty)$ that leads to nonuniqueness requires, in general, that the per capital growth rate of the population be arbitrarily large as x approaches zero. Finite time extinction should result when a low density population is under extreme stress. If this is accomplished by violating uniqueness to solutions of zero initial value problems, the approach can also lead to "creationist" trajectories where nontrivial populations emerge from the initial population $x(t_0) = 0$ on intervals where the population's growth rate is positive.

The mathematical and biological interpretation difficulties arising from finite time extinction are alleviated by employing the concept of ß-extinction.

A population $x = x(t)$ of (1) is said to go to extinction at level ß , $0 \leq ß$, at time T $0 < T \leq \infty$, provided $x(t) > ß$ for $t \in [0, T)$ and $\lim_{t \to T^-} x(t) = ß$; for brevity, this phenomenon is called ß-extinction. A population is ß-persistent on [0, T] if it does not go to ß-extinction for any time in [0, T] .

A main result on ß-survival is

Theorem 4. (i) If, for given $ß > 0$ and for all t in [0, T] , $T < \infty$, the function $G(t, ß) = r(c_0(t)) - ßF(ß)$ is positive then any population with $x(0) > ß$ is ß-persistent on [0, T] .

(ii) If there exists a t_0 in (0, T] where $G(t_0, ß)$ is negative then there is a threshold function, $X = X(t)$, related to certain trajectories associated with (1) which separates these populations that go to ß-extinction at some time in [0, T] from those that do not.

The threshold function in (ii) depends upon g, f, r, and ß ; thus, contrary to the character of the extinction results in Sections 2 and 3, both density dependent and density independent mortality play a role in ß-extinction. However, if ß is small, density dependent mortality is not extremely important from the application viewpoint.

There are analogues of the asymptotic results in Theorems 2 and 3 for fluctuating demographic parameters for ß-persistence and ß-extinction (see Hallam and Ma (1987)).

The theoretical groundwork for ß-survival concepts is recent and ramifications of the work are currently being developed. The concept itself appears suited for application to population-toxicant systems, especially where biological or chemical thresholds are relevant to the modelling process.

6 SURVIVAL IN DISCRETE MODELS

Investigations of the continuous population models above focus upon survival as measured by population growth rate or, in the case of ß-survival, also density dependent effects. Since a growth rate indicates population gradients, additional analysis is needed to ascertain extinction in such models. In a discrete model, the formulation can explicitly involve an extinction function. For example, the difference equation

$$x_{n+1} = x_n[r_n - f(x_n)] , n = 0,1,2,... \tag{4}$$

where $xf(x) > 0$ for $x \neq 0$, is a discrete analogue of the population model (1) . The function

$$G(n, x) \equiv r_n - f(x)$$

no longer represents a growth rate but it is a persistence-extinction function. Persistence ensues at time $n + 1$ if $x_n > 0$ and $G(n, x_n) > 0$ while extinction at time $n + 1$ results when $G(n, x_n) < 0$. Li, Ma, and Hallam (1987) have initiated a survival analysis of (4). The evidence indicates that the extinction behavior of (4) is different from that of (1). This is, of course, no surprise since the behavioral spectrum of an autonomous nonlinear difference equation can be exotic including a chaotic regime (e.g., May (1976)).

It is clear that if $r_n \leq 1$ for all $n \geq 0$ then a population modelled by (4) goes to extinction independent of initial population. The time to extinction is a function of population size and the behavior of f. While it is possible to obtain some results of a general character for (4), to illustrate the extinction phenomena that might arise, the logistic difference equation, where $f(x) = x$,

$$x_{n+1} = x_n(r_n - x_n) , n = 0,1,2,... \tag{5}$$

is discussed.

The survival of a population modelled by (5) is determined iteratively. To demonstrate the procedure and to illustrate the dynamic behavior, suppose $r_2^2 \geq 4r_3$.

Define α_2 and β_2 by

$$a_2 \equiv [r_2 - (r_2^2 - 4r_3)^{1/2}]/2, \ \beta_2 \equiv [r_2 + (r_2 - 4r_3)^{1/2}]/2 ,$$

and let α_i and β_i , $i = 0,1$ be given by

$$\alpha_i = [r_i - (r_i^2 - 4\alpha_{i+1})^{1/2}]/2$$

$$\beta_i = [r_i + (r_i^2 - 4\alpha_{i+1})^{1/2}]/2 .$$

A direct computation using (5) shows that if $r_i^2 \geq 4\alpha_{i+1}$, $i = 0,1$, and if the initial population x_0 satisfies $\alpha_0 \leq x_0 \leq \beta_0$ then x_i satisfies $\alpha_i \leq x_i \leq \beta_i$, $i = 1,2$, and ultimately $x_4 \leq 0$. Also, when $x_2 < \alpha_2$ or $x_2 > \beta_2$ then $x_4 > 0$, so survival occurs for x_2 in these ranges.

An interesting phenomenon arises if $r^2 \geq 4\beta_2$; whenever $x_1 \leq \gamma_1 \equiv [r_1 - (r_1 - 4\beta_2)^{1/2}]/2$ or $x_1 \geq \delta_1 \equiv [r_1 + (r_1 - 4\beta_2)^{1/2}]/2$ it follows that $x_2 \leq \beta_2$. Note $\alpha_1 \leq \gamma_1 \leq \delta_1 \leq \beta_1$. If x_1 is in the intervals $[\alpha_1, \gamma_1]$ or $[\delta_1, \beta_1]$ then, again, extinction occurs at time $n = 4$. However, if x_1 is in $[\gamma_1, \delta_1]$, $x_2 > \beta_2$ and, subsequently, $x_4 > 0$, that is, the solution is persistent at time $n = 4$. Thus, the initial population axis can, in general, be decomposed into intervals where populations go to extinction and intervals where trajectories persist. The general case can be treated in a similar fashion, in that there will exist two intervals or 2^m intervals for some integer m, from which extinction occurs at or before time $N + 1$ while the remainder of the populations are persistent at $t = N + 1$. For our purposes, the variation in $\{r_n\}$ is assumed to result from a variable input in environmental toxicant input. This affects organismal concentration which in turn causes fluctuations in $\{r_n\}$.

From a modelling perspective, the complexity arising in extinction studies of (5) is magnified in population assessment models. In toxicant-population systems often age, size, or stage structure is important because a chemical may affect individuals differently at different stages of the life history, or it may affect individuals of different sizes in distinct ways. A typical model for age or stage structure is the Leslie (1948) or Lefkovitch (1965) matrix model. Generally, each component of the matrix model will contain a formulation based on equations similar to (4) with added interactions, consequently, the extinction behavior can be complicated.

7 SUMMARY

Many current assessment schemes are developed from fate models based upon chemical dynamics with little or no biological content (e.g. EXAMS, Lassiter et al. (1978)) while there are other fate models which include some biological dynamics (Park, et al. (1980); Bartell, et al. (1984)). For several reasons, biological effects have been virtually ignored. First, there is difficulty in determining system effects because the scope of the effects can range from physiological to biospherical. This article, addressing system effects at the population level, indicates a few current developments in survival analysis. Hypotheses that allow temporal variations in demographic parameters are imposed and conclusions obtained contain system level parameters measuring survival of the model population.

Survival in ordinary differential equation models, that is, continuous time population representations, are discussed in some detail. Difference equation models, discrete time population representations, are discussed briefly since theoretical developments have only recently begun in this area.

These results focus on evaluation of population survival through demographic stress and can give only a percussory assessment of the effects of a toxicant on the population. There are many deficiencies of the present theory. To mention only a few, age, stage, or size structure are important features that are omitted. Another factor, lipid content of an individual organism is fundamental for assessing effects for many classes of chemicals (Kooijman (1985)). It is clear that not only the toxic effects of a chemical but the assessment of effects on a population due to exposure to a hazardous chemical is, at present, a risky business but progress is being made.

8 REFERENCES

Bartell, S.M., Gardner, R.H., and O'Neill, R.V., 1984. The Fates of Aromatics Model (FOAM): description, application, and analysis. Ecol. Modelling 22: 109-122.

de Luna, J.T., 1983. Analysis of Mathematical Models of resource-consumer-toxicant interactions. Doctoral dissertaion. University of Tennessee, Knoxville, TN., 140 pp.

Freedman, H.I. and Wa.tman, P., 1984. Persistence in Models of Three Interacting Predator-Prey Populations. Math. Biosciences 68: 213-223.

Freedman, H.I. and Waltman, P., 1985. Persistence in a Model of Three Competitive Populations. Math. Biosciences 73: 89-101.

Hallam, T.G., Clark, C.E., and Jordan, G.S., 1983. Effects of Toxicants on Populations: A Qualitative Approach. II. First Order Kinetics. J. Math. Biol. 18: 25-37.

Hallam, T.G. and de Luna, J.L., 1984. Effects of Toxicants on Populations: A Qualitative Approach. III. Environmental and Food Chain Pathways. J. theor. Biol. 109: 411-429.

Hallam, T.G. and Ma, Zhien, 1986. Persistence in Population Models with Demographic Fluctuations. J. Math. Biology 24: 327-339.

Hallam, T.G. and Ma, Zhien, 1987. On Density and Extinction in Continuous Population Models. J. Math. Biology, to appear.

Hammons, A. (ed), 1981. Ecotoxicology Test Systems. Proceedings of a Series of Workshops. ORNL-5709 EPA 56016-81-004. Oak Ridge National Laboratory, Oak Ridge, TN. 183 pp.

Kooijman, S.A.L.M., 1985. Population Dynamics in Basis of Budgets, TNO Publication P83135a.

Lassiter, R.R., Baughmann, G., and Barnes, L., 1978. Fate of toxic organic substances in the aquatic environment. In S.E. Jorgensen (ed.), State of the Art in Ecological Modelling 7: 211-246. Elsevier, Amsterdam.

Lefkovitch, L.P., 1965. The Study of Population Growth in Organisms Grouped by Stages. Biometrics 21: 1-18.

Leslie, P.H., 1948. Some Further Notes on the use of Matrices in Population Mathematics. Biometrika 35: 213-245.

Li, J., Ma, Zhien, and Hallam, T.G., 1987. Demographic Variation and Survival in Discrete Population Models. IMA J. Math. Med. Biol., to appear.

May, R.M. (1976). Theoretical Ecology. Principles and Applications. Saunders: 317 pp.

Park, R.A., Connolly, C.I., Albanese, J.R., Clesceri, L.S., Heitzman, G.W., Herbrandson, H.H., Indyke, B.H., Loehe, J.R., Ross, S., Sharma, D.D., and Shuster, W.W., 1980. Modelling Transport and behavior of pesticides and other toxic materials in aquatic environments. Report #7. Center for Ecological Modelling, Rensselaer Polytechnic Institute, Troy, NY. 164 pp.

Smith, F.E., 1963. Population Dynamics in Daphnia magna and a New Model for Population Growth. Ecology 44: 651-663.

ON THE GENERAL STRUCTURE OF EPIDEMIC MODELS (*)

E. BERETTA
Istituto di Biomatematica, Universita di Urbino - Via Saffi, 61029 URBINO, ITALY
and
V. CAPASSO
Dipartimento di Matematica, Universita di Bari - Campus - 70125 BARI, ITALY

1 ABSTRACT

The mathematical theory of infectious diseases, as it appears in the literature, is made of various "ad hoc" methods for the analysis of different models.

Here we make an attempt to present a unified approach by means of a general model whose mathematical structure allows a unified treatment as far as the asymptotic behavior of the system is concerned.

The main tool is a suitable Lyapunov functional in order to show global asymptotic stability of equilibrium states.

2 INTRODUCTION

This paper is an attempt to provide a unified approach for the analysis of a large class of epidemic systems.

In Beretta and Capasso (to appear) the authors proposed a general ODE model which includes many of the models proposed up to now by various authors (Anderson and May (1981); Bailey (1975); and Cooke and Yorke (1973)).

The emphasis was devoted to show general theorems for the global asymptotic stability of the nontrivial steady state of the epidemic system, whenever it exists, by means of a Lyapunov functional proposed by Goh (1977, 1978) for a generalized Lotka-Volterra predator-prey system.

In the first part of the present paper the authors discuss the above approach by means of basic models, the SIR models with vital dynamics (see Hethcote (1976)) and the gonorrhea model (Cooke and Yorke (1973)).

Section 5 is devoted to analyze the cases in which the total population is a dynamical variable. The parasite-host system presented in Levin and Pimentel (1981) and an SIS model with vital dynamics proposed in Anderson and May (1981) are studied with respect to global stability of the feasible or partially feasible equilibria.

3 THE GENERAL "ODE" MODEL

We consider first the so called SIR model with vital dynamics (see Hethcote (1976)). If one denotes by S the susceptible population, by I the infective population and by R the removed population, this model may be written as follows:

$$\frac{dS}{dt} = -kSI - \mu S + \mu \,,$$

$$\frac{dI}{dt} = kSI - \mu I - \lambda I \,, \tag{3.1}$$

$$\frac{dR}{dt} = \lambda I - \mu R \,,$$

for $t > 0$, subject to suitable initial conditions. Since clearly (3.1) implies that $S(t) + I(t) + R(t) = 1$ we may ignore the last equation in (3.1) and consider the reduced system

$$\frac{dS}{dt} = -kSI - \mu S + \mu$$

$$\frac{dI}{dt} = kSI - (\mu + \lambda)I \tag{3.2}$$

for $t > 0$.

This model can be written in matrix notations, as follows:

$$\frac{dz}{dt} = \text{diag}(z)\,(e + Az) + c \,, \qquad\qquad t > 0\,, \tag{3.3}$$

if we set $z = (S, I)^T$, and

$$A = \begin{pmatrix} 0 & -k \\ k & 0 \end{pmatrix}, \qquad e = \begin{pmatrix} -\mu \\ -(\mu + \lambda) \end{pmatrix} \,, \qquad c = \begin{pmatrix} \mu \\ 0 \end{pmatrix}. \tag{3.4}$$

For the simple gonorrhea model proposed in Cooke and Yorke (1973), which is an SIS model for two interacting populations (males and females) we have

$$\frac{dI_1}{dt} = -k_1 I_1 I_2 - \alpha_1 I_1 + c_1 k_1 I_2 \,,$$

$$\qquad\qquad , \, t > 0 \,, \tag{3.5}$$

$$\frac{dI_2}{dt} = -k_2 I_1 I_2 - \alpha_2 I_2 + c_2 k_2 I_1 \,,$$

with $c_1 = I_1 + S_1$ and $c_2 = I_2 + S_2$ the two (constant) total populations; S_1 and S_2 denote the two susceptible populations.

Again, (3.5) may be written in matrix notations, as

$$\frac{dz}{dt} = \text{diag}(z)(e + Az) + Bz \qquad , t > 0 , \qquad (3.6)$$

if we set $z = (I_1, I_2)^T$, and

$$A = \begin{pmatrix} 0 & -k_1 \\ -k_2 & 0 \end{pmatrix} , \quad e = \begin{pmatrix} -\alpha_1 \\ -\alpha_2 \end{pmatrix} , \quad B = \begin{pmatrix} 0 & c_1\,k_1 \\ c_2\,k_2 & 0 \end{pmatrix} \qquad (3.7)$$

Altogether, we see that models (3.2) and (3.5) may be considered as particular cases of the following system:

$$\frac{dz}{dt} = \text{diag}(z)\,(e + Az) + b(z) \qquad , \qquad t > 0 \qquad (3.8)$$

with $b(z) = c + Bz$, in $\mathbf{R}^n$, where $n \in \mathbf{N} - \{0\}$ for sake of generality.

Many other epidemic models, proposed by various authors (see e.g. Bailey (1975) and Hethcote (1976)), may be considered as particular cases of (3.8) as shown in Beretta and Capasso (to appear) provided that

(i) $e \in \mathbf{R}^n$ is a real constant vector;

(ii) $A = (a_{ij})_{i,j=1,\dots,\,n}$ is a real constant matrix;

(iii) $b(z) = c + Bz$ is an nonnegative linear vector function defined for $z \in \mathbf{R}^n_+$;

here $c \in \mathbf{R}^n_+$ is a constant nonnegative vector and B is a real constant matrix such that $b_{ij} \geq 0$ for $i, j = 1,\dots, n$ and $b_{ii} = 0$ for $i = 1,\dots, n$.

We may consider two different cases; one in which the total population of the system, defined as $N(t) = \sum\limits_{i=1}^{n} z_i(t)$, is constant, and one in which it is a dynamical variable.

An example of this second class is the following parasite-host system proposed in Levin and Pimentel (1981):

$$\frac{dy}{dt} = (r - k)x - Cxy - Cxv + ry + rv ,$$

$$\frac{dy}{dt} = -(\beta + k)y + Cxy - CSyv \qquad , \quad t > 0 , \qquad (3.9)$$

$$\frac{dv}{dt} = -(\beta + k + \sigma)v + Cxv + CSyv \quad .$$

It can be put again in the matrix form (3.8), if we set

$$e = \begin{pmatrix} r - k \\ -(\beta + k) \\ -(\beta + k + \sigma) \end{pmatrix}, \quad A = \begin{pmatrix} 0 & -C & -C \\ C & 0 & -CS \\ C & CS & 0 \end{pmatrix}, \quad c = 0, \, B = \begin{pmatrix} 0 & r & r \\ 0 & 0 & 0 \\ 0 & 0 & 0 \end{pmatrix}. \tag{3.10}$$

Referring to the general system (3.8) we shall assume at first that either

$$\Omega_1 := \{z \in \mathbf{R}^n_+ | \sum_{i=1}^{n} z_i \leq 1\} \tag{3.11}$$

or

$$\Omega_2 := \{z \in \mathbf{R}^n_+ | z_i \leq 1, \, i = 1,..., n\} \tag{3.12}$$

is positively invariant, as well as their interiors $\overset{\bullet}{\Omega}_1$ or $\overset{\bullet}{\Omega}_2$ respectively. From now on we shall denote both Ω_1 and Ω_2 by Ω.

We shall then make the assumption that

(iv) Ω is positively invariant.

Due to the structure of

$$F(z) := \text{diag}(z)(e + Az) + b(z) \qquad , \, z \in \mathbf{R}^n_+ \tag{3.13}$$

it is clear that $F \in C^1(\Omega)$.

We shall denote by D_i the hyperplane of $\mathbf{R}^n$: $D_i = \{z \in \mathbf{R}^n | z_i = 0|\}$. Clearly $D_i \cap \Omega$ will be positively invariant if $b_{i|D_i} = 0$, while $D_i \cap \Omega$ will be a repulsive set whenever $F(z)$ is pointing inside Ω on D_i, i.e. $b_{i|D_i} > 0$.

Due to the invariance of Ω and the fact that $F \in C^1(\Omega)$, well known fixed point theorems assure the existence of at least one equilibrium solution of (3.8), within Ω.

Our aim here is to show that, under suitable conditions, the general system (3.8) is such that, whenever a strictly positive equilibrium z^* exists in $\Omega_+ := \{z \in \Omega | z_i > 0,$ $i = 1,..., n\}$, then it is globally asymptotically stable with respect to Ω_+. Uniqueness of the equilibrium within Ω_+ clearly follows.

If $z^* = (z^*_1, z^*_2, ..., z^*_n)^T$ is a strictly positive equilibrium of (3.8), then

$$e = -Az^* - \text{diag}(z^{*-1})b(z^*) \tag{3.14}$$

where we have denoted $z^{*-1} := (\dfrac{1}{z^*_1}, ..., \dfrac{1}{z^*_n})^T$.

We may then rewrite (3.8) as follows.

$$\frac{dz}{dt} = \text{diag}(z)\,\tilde{A}\,(z - z^*) - \text{diag}(z - z^*)\,\text{diag}(z^{*-1})b(z) \qquad (3.15)$$

where we have set $\tilde{A} := A + \text{diag}(z^{*-1})B$.

In order to study the qualitative behaviour of system (3.8) we shall refer to the matrix $\tilde{A}$.

We introduce some definitions.

<u>Definition 3.1</u>. Let Q be a real $n \times n$ matrix. We say that $Q \in S_W$ (resp. $Q \in \tilde{S}_W$) iff there exists a positive diagonal real matrix W such that $WB + B^T W$ is positive definite (resp. semidefinite).

<u>Definition 3.2</u>. B is W-skew-symmetrizable (resp. W-symmetrizable) if there exists a positive diagonal real matrix W such that BW is skew-symmetric (resp. symmetric).

As we shall see in Section 4, many epidemic systems are such that either

(v) $\tilde{A}$ is W-skew symmetrizable

or

$$(v')\quad -(\tilde{A} + \text{diag}(-\frac{b_1(z)}{z_1 z_1^*} ,...., -\frac{b_n(z)}{z_n z_n^*})) \in S_W \qquad (v')$$

where $b_i(z)$, $i = 1,..., n$ are the components of the vector function $b(z)$ defined in (iii).

For example, model (3.2) belongs to case (v) while model (3.5) belongs to case (v') .

Under the assumption (v), the elements of $\tilde{A}$ have a skew-symmetric sign distribution. In this case it is useful to associate a graph with $\tilde{A}$ by the following rules:

(A) each component of z , say the i-th is represented by the labeled knot "•" if $b_i(z) = 0$; by ℓ otherwise.

(B) each pair of elements $\tilde{a}_{ij}\,\tilde{a}_{ji} < 0$ is represented by an arc connecting knots "i" and "j" .

By means of the following Lyapunov function proposed in Goh (1977, 1978),

$$V(z) := \sum_{i=1}^{n} w_i(z_i - z_i^* - z_i^* \,\ell n \frac{z_i}{z_i^*}) \qquad (3.16)$$

where $w_i > 0$ are real constants, it is possible to prove the following results (Beretta and Capasso (to appear)).

An account of (3.15), the time derivative of (3.16) along the trajectories of system

(3.8) is given by

$$\dot{V}(z) = (z - z^*)^T \tilde{W}A(z - z^*) - \sum_{i=1}^{n} \frac{w_i \, b_i \,(z)_i}{z_i \, z_i^*}(z_i - z_i^*)^2 =$$

$$= (z - z^*)^T W(\tilde{A} + diag(-\frac{b_1 \,(z)}{z_1 \, z_1^*}, ..., -\frac{b_n(z)}{z_n \, z_n^*}))(z - z^*) \tag{3.17}$$

where W is the positive diagonal matrix $W := diag(w_1, ..., w_n)$. Note that the second expression in (3.17) shows the role of condition (v').

Lemma 3.1. Assume that $\tilde{A}$ is W-skew symmetrizable. If the associated graph is either (a) a tree and $\rho - 1$ of the ρ terminal knots are ℓ, or (b) a chain and two consecutive internal knots are ℓ, or (c) a cycle and two consecutive knots are ℓ, then $M \equiv \{z^*\}$ within R.

Theorem 3.1. Suppose that system (3.8) has a positive equilibrium $z^* \in \Omega_+$, and case (v) holds true under one of the hypotheses of Lemma 3.1, or otherwise suppose that (v') holds true; then the positive equilibrium z^* is globally asymptotically stable within Ω_+.

Corollary 3.1. If the vector c in (iii) is positive definite, then system (3.8) has a positive equilibrium $z^* \in \Omega_+$. In case (v), and under the assumptions of Lemmas 3.1, or (v') the positive equilibrium z^* is globally asymptotically stable with respect to Ω_+.

Remark 3.1. Under the assumptions of Theorem 3.1 or Corollary 3.1 the uniqueness of z in Ω_+ follows from its global asymptotic stability.

In the case in which the total population of the system $N(t) = \sum_{i=1}^{n} z_i(t)$ is a dynamical variable we must drop assumption (iv).

For these models, the accessible space is the whole non-negative orthant R_+^n of the Euclidean space, and Theorem 3.1 can be reformulated by substituting the bounded set Ω_+^n with the positive orthant R_+^{n*}. Furthermore, it has to be noticed that we cannot directly apply fixed point theorems even when vector c is positive definite. Hence, Corollary 3.1 cannot be applied to these models. However, the structure of system (3.8) is such that the positive invariance of the non-negative orthant R_+^n is assured. When vector c in (iii) has some identically vanishing components, system

(3.8) may have equilibria z^* with some vanishing components.

Let N be the set of indices $N = \{1,..., n\}$ and I the subset of N such that $z_i = 0$ when $i \in I$. According to Goh (1977) if $I \neq \emptyset$ then z is partially feasible and we can study the sectorial stability of z^* with respect to

$$\mathbf{R}^n_I = \{z \in \mathbf{R}^n_+ | z_i > 0 , i \in N - I\} . \tag{3.18}$$

The definition of sectorial stability can be found in Hethcote (1976). By sectorial stability we mean that z is globally asymptotically stable with respect to $\mathbf{R}^n_+$. If $I = \emptyset$ then z is feasible and we can study its global asymptotic stability with respect to $\mathbf{R}^*_+$ by Theorem 3.1.

Assume that z^* is a partially feasible equilibrium of (3.8), i.e., $z^*_i = 0 , i \in I, I \neq \emptyset$. Define the matrix $\tilde{A} = (\tilde{a}_{ij})_{i,j=1,...,n}$ as:

$$\tilde{a}_{ij} = a_{ij} + \frac{b_{ij}}{z^*_i} , \qquad \text{for all } i \in N - I, j \in N ,$$

$$\tag{3.19}$$

$$\tilde{a}_{ij} = a_{ij} , \qquad \text{otherwise,}$$

where a_{ij} are the elements of the matrix A in (3.8) and b_{ij} are the elements of B defined in (iii). Let R be the subset of $\mathbf{R}^n_I$ such that:

$$R = \{z \in \mathbf{R}^n_I | z_i = 0 \text{ for all } i \in I , z_i = z^*_i \text{ for all } i \in N - I \text{ s.t. } b_i(z) > 0\} \tag{3.20}$$

and let M be the largest invariant set within R. By the scalar function suggested by Goh in studying sectorial stability

$$V(z) = \sum_{i \in N-I} w_i[z_i - z^*_i - z^*_i \ \ell n \ \frac{z_i}{z^*_i}] + \sum_{i \in I} w_i z_i , \qquad w_i > 0 ,$$

it is possible to prove the following (Beretta and Capasso (to appear)):

Theorem 3.2. Let z^* be a partially feasible equilibrium of (3.8) and assume that $\tilde{A}$ is W-skew symmetrizable. If (a) $e_i + \sum_{j \in N} a_{ij} z^* \leq 0$ for all $i \in I$, (b) $b_i(z) \equiv 0$ for all $i \in I$, (c) $M \equiv \{z^*\}$, then z^* is globally asymptotically stable with respect to $\mathbf{R}^n_+$.

From Theorem 2.2 we obtain:

<u>Corollary 3.2</u>. Let z^* be a feasible equilibrium of (3.8) and assume that $\tilde{A}$ is W-skew symmetrizable. if $M \equiv \{z^*\}$, then z^* is globally asymptotically stable with $R_+^{n^*}$.

Corollary 3.2 can be seen as a new formulation of Theorem 3.1. case (v), for epidemic models for which assumption (iv) is dropped. Concerning Corollary 3.2. we may observe that, if the graph associated with $\tilde{A}$ by the rules (vii), (viii), satisfies one of the hypotheses of Lemma 3.1., then within R we have $M \equiv \{z^*\}$.

4 APPLICATIONS
4.1. <u>SIR model with vital dynamics</u>
If we refer to model (3.2), since in this case $B = 0$, we have

$$\tilde{A} = A = \begin{pmatrix} 0 & -k \\ k & 0 \end{pmatrix} \quad , \quad b(z) = c = \begin{pmatrix} \mu \\ 0 \end{pmatrix} . \tag{4.1}$$

Since $\tilde{A}$ is skew symmetric and the associated graph is $\ell \longrightarrow$, Theorem 3.1 applies.

In this case the equilibrium point is given by

$$S^* = \frac{\mu + \gamma}{k} \quad , \quad I^* = \frac{\mu}{k} \frac{(1 - S^*)}{S^*} \quad , \quad R^* = \frac{\gamma}{k} \frac{(1 - S^*)}{S^*} \tag{4.2}$$

provided that $S^* < 1$.

4.2. <u>Gonorrhea model</u>
Consider the epidemic model (3.5) and assume for simplicity that $k_1 = k_2 = 1$. Then the model can be rewritten as

$$\frac{dI_1}{dt} = -I_1 I_2 - \alpha_1 I_1 + c_1 I_2 , \qquad I_1 + S_1 = c_1 ,$$

$$\tag{4.3}$$

$$\frac{dI_2}{dt} = -I_1 I_2 - \alpha_2 I_2 + c_2 I_1 , \qquad I_2 + S_2 = c_2 .$$

In this case

$$b(z) \equiv Bz = (c_1 l_2, c_2 l_1)^T ; \quad \tilde{A} = \begin{pmatrix} 0 & \dfrac{c_1 - l_2^*}{l_1^*} \\ \dfrac{c_2 - l_2^*}{l_2^*} & 0 \end{pmatrix}$$

hence

$$W[\tilde{A} + \text{diag}(-\dfrac{c_1 l_2}{l_1^* l_1}, -\dfrac{c_2 l_1}{l_2^* l_2})] = \begin{pmatrix} -\dfrac{c_1 l_2}{l_1^* l_1} w_1 & w_1 \dfrac{S_1^*}{l_1^*} \\ w_2 \dfrac{S_2^*}{l_2^*} & -\dfrac{c_2 l_1}{l_2^* l_2} w_2 \end{pmatrix} \qquad (4.4)$$

which is a symmetric matrix if we choose $w_1 > 0$ and w_2 such that

$$w_2(S_2^* / l_2^*) = w_1 (S_1^* / l_1^*) .$$

The symmetric matrix (4.4) is negative definite. In fact, the diagonal elements are negative and

$$(\dfrac{c_1 l_2}{l_1^* l_1} \dfrac{c_2 l_1}{l_2^* l_2} - \dfrac{S_1^* S_2^*}{l_1^* l_2^*}) w_1 w_2 = \dfrac{w_1 w_2}{l_1^* l_2^*} (c_1 c_2 - S_1^* S_2^*) > 0 , \qquad (4.5)$$

where the fact that $0 < S_i^* < c_i$ is taken into account since z^* is a positive equilibrium.

Since $W[\tilde{A} + \text{diag}(-(c_1 l_2 / l_1^* l_1), -(c_2 l_1 / l_2^* l_2))]$ is symmetric and negative definite,

then $-[\tilde{A} + \text{diag}(-(c_1 l_2 / l_1^* l_1), -(c_2 l_1 / l_2^* l_2))] \in S_W$.

5 MODELS WITH DYNAMIC TOTAL POPULATION

5.1 Parasite-host system

We discuss now the model (3.9) proposed in Levin and Pimentel (1981).

As discussed in Levin and Pimentel (1981) the two cases $r < k$ and $r > \beta + k + \sigma$ do not give rise to nontrivial equilibrium solutions. We shall then restrict our analysis to the case $\beta + \sigma + k > r > k$ in which there is an equilibrium at

$$x^* = \frac{r}{C}\,\frac{\sigma}{\sigma - S(r-k)}\ ,\quad y^* = \frac{\beta+k+\sigma}{CS} - \frac{1}{S}x^*,\quad v^* = \frac{1}{S}x^* - \frac{\beta+k}{CS}\ . \tag{5.1}$$

Local stability results were already given in Levin and Pimentel (1981). According to the aim of this paper we shall study global asymptotic stability of the feasible or partially feasible equilibrium.

The equilibrium $z^* = (x^*, y^*, v^*)^T$ is feasible, i.e., its components are positive if

$$\frac{r}{\beta+k+\sigma} < 1 - \frac{S(r-k)}{\sigma} < \frac{r}{\beta+k}\ . \tag{5.2}$$

If $\sigma < \sigma_1$ where σ_1 is such that

$$\frac{r}{\beta+k+\sigma_1} = 1 - \frac{S(r-k)}{\sigma_1}\ , \tag{5.3}$$

the first inequality in (5.2) is violated and only a partially feasible equilibrium is present given by

$$x^* = \frac{\beta+k+\sigma}{C}\ ,\quad y^* = 0\ ,\quad v^* = \frac{r-k}{\beta+k+\sigma-r}\,x^*\ , \tag{5.4}$$

since $r < \beta + k + \sigma$. If $\sigma = \sigma_1$ then (5.1) coalesces in (5.4).

If $r < \beta + k$ and $\sigma > \sigma_2$, where σ_2 is such that

$$1 - \frac{S(r-k)}{\sigma_2} = \frac{r}{\beta+k}\ , \tag{5.5}$$

then the second inequality in (5.2) is violated and only a partially feasible equilibrium is present, given by

$$x^* = \frac{\beta+k}{C}\ ,\quad y^* = \frac{r-k}{\beta+k-r}\,x^*\ ,\quad v^* = 0\ , \tag{5.6}$$

since $r > k$. If $\sigma = \sigma_2$ then (5.1) coalesces in (5.6).

Now consider the case in which the equilibrium (5.1) is feasible, i.e., $z^* \in R^*_+$. Then by using the notations of Section 3

$$b(z) \equiv Bz\ ,\qquad \tilde{A} = \begin{pmatrix} 0 & -(C - \frac{r}{x^*}) & -(C - \frac{r}{x^*}) \\ C & 0 & -CS \\ C & CS & 0 \end{pmatrix} \tag{5.7}$$

where z is a vector $z = (x, y, v)^T$ belonging to the non-negative orthant R_+^3. Since $C - r/x^* = C (S(r - k))/\sigma$, provided that $r > k$, matrix $\tilde{A}$ is W-skew symmetrizable by the diagonal positive matrix $W = \text{diag}(w_1, w_2, w_3)$ where $w_1 = \sigma/S(r - k)$, $w_2 = w_3 = 1$. In fact we obtain:

$$W\tilde{A} = \begin{pmatrix} 0 & -C & -C \\ C & 0 & -CS \\ C & CS & 0 \end{pmatrix} \tag{5.8}$$

Now, we are in position to apply Corollary 3.2. Since $b(z) = (r(y + z), 0, 0)^T$, the subset of all points within R_+^{3*} where we have $\dot{V}(z) = 0$, is:

$$R = \{z \in R_+^3 | x = x^*\} . \tag{5.9}$$

Now, we look for the largest invariant subset M within R. Since $x = x^*$ for all t, $dx/dt|_R = 0$, and from the first of the equations (3.9) we obtain:

$$(y + v)|_R = (r - k)/(C - r/x^*)) = (\sigma/CS) \quad \text{for all } t. \tag{5.10}$$

Therefore $\dfrac{d(y + v)}{dt}|_R = 0$, and by the last two equations (3.9) we obtain:

$$v|_R = \frac{1}{\sigma} \{[Cx^* - (\beta + k)] [(y + v)]_R\} = \frac{1}{CS} [Cx^* - (\beta + k)] = \frac{x^*}{S} - \frac{\beta + k}{CS} . \tag{5.11}$$

Then taking into account (5.1) we have $z|_R \equiv z^*$. It immediately follows that

$$y|_R = \frac{\sigma}{CS} - v^* = \frac{\beta + k - \sigma}{CS} - \frac{x^*}{S} , \tag{5.12}$$

i.e., $y|_R = y^*$. Then the largest invariant set M within R is z^*. From Corollary 3.2 the global asymptotic stability of the feasible equilibrium (5.1) within R_+^{3*} follows.

It is to be noticed that the only assumptions made in this proof are $r > k$ and that equilibrium (5.1) is feasible. Under these assumptions we exclude that unbounded solutions may exist.

Suppose that $\sigma \leq \sigma_1$, i.e., the equilibrium (5.1) is not feasible and we get the partially feasible equilibrium (5.4) which belongs to:

$$R^3_{\{2\}} = \{z \in R^3_+ | z_i > 0 , i = 1, 3\} . \tag{5.13}$$

In order to apply Theorem 3.2 hypotheses (a), (b) must be verified.
Hypothesis (a) requires:

$$-(\beta + k) + cx^* - cS\, v^* \leq 0 , \tag{5.14}$$

from which, taking into account (5.4), we obtain:

$$1 - \frac{S(r - k)}{\sigma} \leq \frac{r}{\beta + k + \sigma} . \tag{5.15}$$

Inequality (5.15) is satisfied in the whole range $\sigma \leq \sigma_1$, within which the partially feasible equilibrium (5.14) occurs. When $\sigma = \sigma_1$, the equality applies in (5.15). Hypothesis (b) is satisfied because $b(z) = (r(y + v), 0, 0)^T$ and therefore $b_2(z) \equiv 0$. Concerning hypothesis (c), consider first the case $\sigma < \sigma_1$, i.e., the inequality applies in (5.14).
Then, the subset (3.20) is:

$$R = \{z \in R^3_{\{2\}} | y = 0 ; x = x^*\} . \tag{5.16}$$

Now, we look for the largest invariant subset M within R .
Since $x = x^*$, $y = 0$ for all t , $dx/dt|_R = 0$, and from the first of equations (3.9) we get

$$v|_R = (r - k)/(C - r/x^*) , \quad \text{where } x^* = \frac{\beta + k + \sigma}{C} . \tag{5.17}$$

Therefore, we obtain: $v|_R = (r - k)/(\beta + k + \sigma - r)\, x^*$, i.e., $v|_R \equiv v^*$. Thus, the largest invariant set within R is:

$$z^* = (x^* = \frac{\beta + k + \sigma}{C} , y^* = 0 , v^* = \frac{r - k}{\beta + k + \sigma - r}\, x^*)^T . \tag{5.18}$$

When $\sigma = \sigma_1$, then equality applies in (5.15) and (3.20) becomes:

$$R = \{z \in R^3_2 | x = x^*\} .$$

In this case, we have already proved that $M \equiv \{z^*\}$. Hence, hypothesis (c) is satisfied. Then, by Theorem 3.2 the partially feasible equilibrium is globally asymptotically stable with respect to $\mathbf{R}^3_{\{2\}}$.

If $r < \beta + k$ and $\sigma \geq \sigma_2$, then the partially feasible equilibrium (5.4) occurs. This equilibrium belongs to

$$\mathbf{R}^3_{\{3\}} = \{z \in \mathbf{R}^3 | z_i > 0 , i = 1,2 ; z_i \geq 0 , i = 3\} . \tag{5.19}$$

Hypothesis (a) of Theorem 3.2 requires

$$-(\beta + k + \sigma) + Cx^* + CSy^* \leq 0 , \tag{5.20}$$

from which, taking into account (4.11), we obtain:

$$1 - \frac{S(r - k)}{\sigma} \geq \frac{r}{\beta + k} . \tag{5.21}$$

This inequality is satisfied in the whole range of existence of the equilibrium (5.6), i.e., for all $\sigma \geq \sigma_2$. When $\sigma = \sigma_2$, the equality applies in (5.21). Hypothesis (b) of Theorem 3.2 is obviously satisfied. Concerning hypothesis (c), at first we consider the case in which $\sigma > \sigma_2$. Therefore the inequality applies in (5.20) and the subset (3.20) of $\mathbf{R}^3_3$ is:

$$R = \{z \in \mathbf{R}^3_3 | v = 0 , x = x^*\} . \tag{5.22}$$

From (5.6), we are ready to prove that $M \equiv \{z^*\}$. When $\sigma = \sigma_2$, R becomes

$$R = \{z \in \mathbf{R}^3_3 | x = x^*\}$$

and we have already proved that $M \equiv \{z^*\}$. Hypothesis (c) is satisfied.

Also in this case Theorem 3.2 assures the global asymptotic stability of the partially feasible equilibrium (5.16) with respect to $\mathbf{R}^3_3$.

5.2 SIS Model with Vital Dynamics (Anderson and May (1981))

$$dS/dt = (r - b)S - \rho SI + (\mu + r)I \tag{5.23}$$

$$dI/dt = -(\theta + b + \mu)I + rSI ,$$

where, denoting by $n = S + I$, we have

$$dn/dt = (r - b)n - \theta I . \tag{5.24}$$

Provided that $r > b$, $\theta > r - b$, system (5.24) has the feasible equilibrium $z^* \in R_+^{2*}$:

$$S^* = \frac{\theta + b + \mu}{\rho} , \quad 1^* = \frac{r - b}{\theta + b - r} S^* . \tag{5.25}$$

When $r \leq b$, or $r > \theta + b$, the equilibrium (5.25) is not feasible and the only equilibrium of (5.23) is the origin. System (5.23) may be put in the form (3.8), where

$$e = \begin{pmatrix} r - b \\ -(\theta + b + \mu) \end{pmatrix} , \quad A = \begin{pmatrix} 0 & -\rho \\ \theta & 0 \end{pmatrix} , \quad B = \begin{pmatrix} 0 & \mu + r \\ 0 & 0 \end{pmatrix} , \quad C = 0 \tag{5.26}$$

and $b(z) \equiv Bz = ((\mu + r)I, 0)^T$. When z^* is a feasible equilibrium the matrix $\tilde{A} = A + \mathrm{diag}(z^{*-1})B$ is given by

$$A = \begin{pmatrix} 0 & -\left[\rho - \frac{(\mu + r)}{S^*}\right] \\ \rho & 0 \end{pmatrix} \tag{5.27}$$

Since $S^* = (\theta + b + \mu)/\rho$, provided that $\theta > r - b$ the matrix $\tilde{A}$ is skew-symmetrizable. Because $b_1(z) \geq 0$, the graph associated with $\hat{A}$ is •—• and by Corollary 3.2 the global asymptotic stability of z^* with respect to R_+^2, follows.

When $r \leq b$, $r > \theta + b$ Theorem 3.2 cannot be applied to study attractivity of the origin because hypothesis (b) is violated.

6 REFERENCES

Anderson, R.M. and May, R.M., 1981. The population dynamics of micro-
    parasites and their invertebrate hosts. Trans. Roy. Phil. Soc. (B), 291: 451-524.
Bailey, N.T.J., 1975. The Mathematical Theory of Infectious Diseases, Griffin, London.
Beretta, E., and Capasso, V., to appear. On the general structure of epidemic systems.
    Global asymptotic stability.
Cooke, K.I, and Yorke, J.A., 1973. Some equations modelling growth processes and
    gonorrhea epidemics. Math. Biosci. 16: 75-101.
Goh, B.S. 1977. Global stability in many species systems. Am. Nat., 111: 135-143.
_____, 1978. Global stability in a class of prey-predator models. Bull. Math. Biol., 40:
    525-533.
Hethcote, H.W., 1976. Qualitative analyses of communicable disease models.
    Math Biosci., 28: 335-356.
Levin, S., and Pimentel, D., 1981. Selection of intermediate rates of increase in parasite-
    host systems. Am. Nat., 117: 308-315.
Takeuchi, Y., Adachi, N., and Tokumaru, T., 1978. The stability of generalized Volterra
    equations. J. Math. Anal. Appl. 62: 453-473.

EQUILIBRIA AND OSCILLATIONS IN AGE-STRUCTURED POPULATION GROWTH MODELS

J.M. CUSHING
Department of Mathematics, University of Arizona,Tucson, Arizona U.S.A.

1 INTRODUCTION

The purpose of this paper is to describe some recent results concerning the dynamics of some quite general age-structured population growth models and hopefully to demonstrate the usefulness of a certain approach to the study of asymptotic states. This approach is based upon bifurcation theory and methods. Its benefits lie not only in the resulting wealth of mathematical techniques which can be brought to bear on the equations, both for the detailed study of specific model equations and for the derivation of very general and generic results for more general models, but also in the conceptional simplification, organization and unification of many phenomena which appear in common to a wide diversity of models and model equations types.

The model equations to be studied here are sufficiently general as to include virtually all equations for single population growth dynamics and for many species interactions which appear in the literature, whether they involve ordinary differential equations, delay or functional differential equations, integrodifferential equations, integral equations, or (by discretization) difference and Leslie matrix-type equations.

The motivation for the approach lies in the observation that if the asymptotic states are studied as a function of a model parameter measuring an "inherent growth" or "birth" rate, here denoted by n , then the following scenario usually occurs: for "small" values of n the population does not survive in that its density tends to zero as $t \to \infty$ and only if n surpasses a critical value can the population survive in the sense that there exists a stable equilibrium state (not necessarily unique). Furthermore for larger values of n the equilibrium sometimes destabilizes and a sustained time periodic oscillation results. Moreover, further increases in n can sometimes result in repeated destabilizations and even "chaotic" dynamics as is so familiar now with well studied difference equations for populations with nonoverlapping generalizations.

The model assumes that each population can be described by a density $w = w(t, x) \geq 0$ where t is time and x is a characteristic which structures the population (such as age, size, length, weight, amount of some chemical, etc.) whose growth rate is given by $dx/dt = v$. In general v depends on t, x, and w . The dynamics of w are governed by the equation

$$d_t w + d_x(vw) + \delta = 0 \, , t > 0 \, , 0 < x < X \leq + \infty$$

where δ is a growth rate depending in general on t, x, and w (Nisbet and Gurney (1982); Sinko and Streifer (1967)) . Side conditions, besides the prescription of an initial condition $w(0, x)$, include $w(t, x) \equiv 0$ for $x \geq X$ (X is the maximal characteristic value for all members of the population) and a birth law $vw = \beta$ at $x = 0$ where the birth rate β is a function of t and w. Here it is assumed that all newborns have the same characteristic value which has been normalized to $x = 0$.

The mathematical theory of these model equations is not well developed, at least in this generality, and even such fundamental mathematical questions as the existence and uniqueness of solutions await study. One case which in recent years has been well studied is the case when v is independent of the density w and t, i.e. when $v = v(x) > 0$. For this case a change of variables from x and w to the variables

$a = \int_0^x (1/v(s))\, ds$ and $u = vw$ reduces the equations to similar equations for the density

$u = u(t, a)$ in which $v \equiv 1$. In this equivalent formulation the population is structured by age a and it is with these equations with which I will deal in this paper.

I will assume here that there is no seeding/harvesting or immigration/emigration so that the vital rates δ, β drop to zero when the density u is zero. Furthermore, I will assume, although it is not mathematically necessary for many of the results to be described below, that δ is a death rate, i.e., $\delta \leq 0$ (which is consistent with the only entries into the population being births). Actually, for simplicity of presentation, I will assume here that the vital birth and death rates β and δ can be expressed in terms of nonnegative per unit density rates F and D. Thus, in summary, I will consider in this paper model equations of the form

(a) $d_t u + d_a u + u D = 0$, $t > 0$, $0 < a < A \leq +\infty$

(b) $u(t, 0) = \int_0^A u F\, da$, $t > 0$ (1)

(c) $u(t, a) = 0$, $t > 0$, $a \geq A$

where A is the maximal possible age and D, F depend in general explicitly on t, a and (in so-called density dependent models) on u. Rather than write $D = D(t, a, u)$, $F = F(t, a, u)$ however, I will for the sake of simplicity write $D = D(u) \geq 0$, $F = F(u) \geq 0$ in which the dependence on the independent variables t, a is notationally surpressed.

These equations can be found in early works of McKendrick (1926) and in later works of several authors (Hoppenstaedt (1975); Sinko and Streifer (1967); von Forester (1959)). A seminal paper on nonlinear versions of (1) which initiated a great deal of research interest during the last ten years is that of Gurtin and MacCamy (1974). A recent modern mathematical treatment can be found in the new book by Webb (1985) (which also has a large bibliography). Also forthcoming is a book by Gurtin (to appear) on age-structured population dynamics.

The modelling of a specific population growth problem is done by assigning specific properties to the vital per unit rates D and F. Often the resulting equations can be manipulated into other simpler equations. For example if these vital rates are not age specific and if they depend on density u only through total population size

$P = \int_0^A u(t, a)\, da$ then an integration of (1a) from 0 to A leads to the equation

$d_t P = PG(t, P)$ where $G = F - D$ is a per unit net growth rate. Thus, classical ordinary differential equation models (of so-called Kolmogorov type) for total population size are included in (1) as special cases. Often other special cases also lead to ordinary differential equation models (Gurtin (to appear); Gurtin and Levine (1979)). To see how integrodifferential and delay functional differential equations can be derived from (1) (see Cushing (1980 and 1981)). Renewal Voltera type integral equations and difference equations can also arise from (1) (see (Cushing (1981) and Oster (1977)).

In this paper the emphasis will be on the general model (1) with as few restrictive assumptions as possible. It will be seen that a great deal can be said about some fundamental properties of the long time asymptotic states under only the mildest of assumptions on the vital death and birth rates D and F as they depend on population density in an arbitrary way.

First equilibria for single species models will be considered. Then equilibria for m interacting age-structured populations will be considered using similar techniques. Finally, some brief remarks will be made concerning unforced and forced periodicities in population densities.

2 SINGLE SPECIES GROWTH MODELS
2.1. A Global Existence Result.

If the vital rates D and F do not explicity depend on time t then the first order hyperbolic partial differential equation (1a) is autonomous and an important role is played by equilibrium solutions $u = u(a) \geq 0$. The equilibrium equations are

$$d_a u + uD = 0 \,,\, 0 < a < A$$
$$u(0) = \int_0^A uF\, da \,,\, u(a) = 0 \text{ for } a \geq A \,. \tag{2}$$

First, without loss in generality, write $D = \mu + d(u)$ where $d(0) = 0$. Now suppose that $\mu = \mu(a)$ is a continuous real-valued function on $[0, A)$ which satisfies $M(A-) = +\infty$ where $M(a) := \int_0^a \mu\, ds$ and let $u_0(a) := \exp(-M(a))$ for $0 \leq a < A$ and 0 for $a \geq A$. $u_0(a)$ is the probabiliity of living to age a.

The function u_0 is continuous for $a \geq 0$ and can be used to define a Banach space $B = B(\mu)$ of continuous functions $u: R \to R$, with support on $[0, A]$, for which u/u_0 is continuous on $[0, A]$ under the norm $\sup_{0 \leq a \leq A}|u/u_0|$. A solution of (2) in B is a differentiable function on $(0, A)$. Note that solutions in B automatically satisfy the condition $u(a) = 0$ for $a \geq A$. A positive solution in B satisfies $u > 0$ on $[0, A)$. Note that (2) has the trivial solution $u \equiv 0$ in B. Also note that $u_0 \in B$ solves $d_a u + \mu u = 0$.

Suppose that $F = nf$ where $n \in R$ and $f = f(u)$ is normalized so that

$$\int_0^A u_0 f(0) \, da = 1 . \tag{3}$$

Then $n = \int_0^A u_0 F(0) \, DA$ is the inherent net reproductive rate, i.e., the expected number of offspring per individual per lifetime.

Question: for what values of the inherent net reproductive rate n do the equilibrium equations in (2) have a positive solution in B ? and when it is stable as a solution of the dynamical equations (1)?

An equilibrium pair $(n, u) \in R \times B$ is a pair for which u is a solution of (2) in B with $F = nf$. A positive equilibrium pair is one for which u is a positive equilibirum solution. Equations (2) have the trivial solution pairs $(n, 0)$ for all $n \in R$.

A fundamental principle of bifurcation theory is that bifurcation from the trivial solution $u \equiv 0$ can occur only at a value of n at which the linearized equations have nontrivial solutions. It is elementary to see that the linear equations

$$d_a u + \mu u = 0 \, , \, u(0) = n \int_0^A u f(0) \, da$$

have a nontrivial solution in B if and only if $n = 1$ (cf. (3)) in which case all solutions are constant multiples of u_0. Thus the only possible bifurcation point is $(n, u) = (1, 0)$. But does bifurcation of positive equilibrium pairs actually occur at $(1, 0)$?

Assume $F = nf$, $f = \phi + r(u)$ where $r(0) \sim 0$ and $\phi = f(0) \in L_1[0, A]$ satisfies the normalization (3). Suppose $A < +\infty$. If the higher order operators

$$\int_0^A ur(u) \, da \, , \, ud(u) \tag{4}$$

are continuous as operators from B into R and B respectively, then it is possible to show [6], using abstract bifurcation theory methods, that the equilibrium equations (2) possess a continuum C (i.e. a closed and connected set) of equilibrium pairs (n, u) with the following properties:

$(0, 1) \in C$

$(n, u) \in C/(1, 0)$ is a positive equilibrium pair

C is unbounded in $R \times B$.

This fundamental existence result shows that the primary bifurcation in the scenario described in the Introduction is, with regard to the inherent net reproductive rate of the population and positive equilibrium states, a universal property of population growth models. Note the biological significance of the critical value $n = 1$, namely when $n = 1$ birth and death are balanced so that exact per unit replacement occurs at low (technically zero) density.

It must be kept in mind that for specific model equations it is possible that positive equilibrium pairs exist which do not lie on the bifurcating continuum C. For an example see Cushing (1984).

The higher order terms (4) need not have the form above, i.e. δ, β need not arise from per unit density vital rates D and F, for this general existence result although in the general case only positivity near the bifurcation point $(1, 0)$ can be guaranteed (see Cushing (1985)).

If $A = +\infty$ one loses a crucial compactness property in the proof of the above result and this global result (i.e. the unboundedness of C) remains an open question in this case. Local bifurcation still occurs however (see Cushing (1984)).

2.2. Local Analysis.

One wishes to know not only the existence of positive equilibrium densities, but also about the stability properties of these equilibria as solutions of the dynamical equations (1). In general one cannot expect that the equilibrium pairs on the continuum C are stable. Indeed, as already mentioned in the Introduction, even if the pairs "start out" stable near the bifurcation point $(1, 0)$ they need not globally remain stable. For example, Hopf-type bifurcation from equilibria to nontrivial time periodic solutions of (1) is a possibility. Moreover, the positive equilibria from C need not "start out" stable near $(1, 0)$ as will be seen.

Locally, near the bifurcation point, it is possible to prove some precise and very general results concerning the (linearized) stability of both the trivial equilibrium $u \equiv 0$ and the positive equilibria from the continuum C. First, a refinement of the existence result above can be made near $(1, 0)$ by means of classical implicit function theorem (or Liapunov-Schmidt) methods. Suppose $A \leq +\infty$ and that the operators in (4) are $q \geq 1$ times continuously (Fréchet) differentiable near $u = 0$. Then the (necessarily unique) positive equilibrium pairs near the bifurcation point can be expressed for $\varepsilon > 0$ sufficiently small as

$$n = n(\varepsilon) = 1 + \gamma(\varepsilon) , u = u(\varepsilon) = \varepsilon u_0 + \varepsilon u_1(\varepsilon) \tag{5}$$

where γ, u_1 are continuously differentiable as functions of ε for small $|\varepsilon|$ near $\varepsilon = 0$ which map into R and B_0 (the functions u in B satisfying $\int_0^A uu_0 \, da = 0$) and which

satisfy $\gamma(0) = 0 , u_1(0) = 0$.

In order to study the stability of both these positive equilibrium solutions and the trivial equilibrium $u \equiv 0$, the dynamical equations can be linearized at the equilibrium in question and solutions of the resulting linear equations sought in the form $u = y \exp(zt)$ where $0 \neq y \in B$ and z is a complex number. If such a solution can be found with $\mathrm{Re}\, z > 0$ then the equilibrium will be called <u>unstable</u>. If no such solutions exist for $\mathrm{Re}\, z \geq 0$ then the equilibrium will be called <u>stable</u> (locally asymptotically stable). For a rigorous justification of this standard procedure for (1) (although in a slightly different setting) see Webb (1985).

The stability of the trivial pairs $(n, u) = (n, 0)$ is easily ascertained by this method, for the linearization of (1) at $u = 0$ with $F = nf$ leads, with $u = y \exp(zt)$, to the equations

$$d_a y + y(\mu + z) = 0, \quad y(0) = n \int_0^A y\, \phi\, da.$$

This pair of equations is easily seen to have a nontrivial solution $y \in B$ if and only if z satisfies the "characteristic equation"

$$1 = nc(z), \quad c(z) := \int_0^A e^{-za}\, u_0\, \phi\, da.$$

Inasmuch as $|nc(z)| < 1$ for $n < 1$ and $\mathrm{Re}\, z \geq 0$ while for $n > 1$, $nc(0) > 1$ and $nc(z) \to 0$ as $z = \mathrm{real} \to +\infty$ it is seen that $u = 0$ is stable for $n < 1$ and unstable for $n > 1$.

<u>Thus the trivial equilibrium $u \equiv 0$ loses stability as the inherent net reproductive rate</u> n <u>increases through the critical value</u> 1.

The stability of the positive bifurcating equilibria (5) is more difficult to study. The linearization procedure leads to linear homogeneous equations in which the parameter ε appears. It is possible to show (Cushing (1985 and 1984)) that these equations possess "eigen-solutions" of the form $u = y(\varepsilon) \exp(z(\varepsilon)t)$ for $|\varepsilon|$ small where $y(\varepsilon) = u_0 + v(\varepsilon)$, $z = z(\varepsilon)$, with $v(0) = 0$, $z(0) = 0$. Moreover

$$z'(0) = -\gamma'(0)/m_1, \quad m_1 := \int_0^A a u_0(a)\, \phi(a)\, da > 0. \tag{6}$$

Thus $z(\varepsilon) > 0$ if $\gamma'(0) < 0$ and $z(\varepsilon) < 0$ if $\gamma'(0) > 0$ for $\varepsilon > 0$ small.

Since $\gamma'(0) < 0$ means $n < 1$ for $\varepsilon > 0$ small this case is called <u>subcritical bifurcation</u>. Similarly the case $\gamma'(0) > 0$ is called <u>supercritical bifurcation</u> and corresponds to $n > 1$. The result (6) implies that the stability of the bifurcating branch depends on the direction of bifurcation.

<u>Thus, locally near bifurcation, the positive equilibrium pairs (5) from</u> C <u>are stable if and only if the bifurcation supercritical or in other words</u> $n > 1$.

The implicit function theorem methods used to obtain this result also yield a formula for $\gamma'(0)$:

$$\gamma'(0) = \int_0^A u_0 \phi \int_0^a d'(0) (u_0) \, ds \, da - \int_0^A \phi r'(0) (u_0) da \tag{7}$$

where $d'(0)$ and $r'(0)$ are Fréchet derivatives at $u = 0$. Thus the direction of bifurcation is determined by the (age-specific) changes in the vital rates D and f with respect to density u near $u = 0$.

For example, if an increase in density (at low levels) is deleterious in the sense that the death rate cannot decrease and the birth rate cannot increase, i.e. $d'(0)(u_0) \geq 0$, $f'(0)(u_0) \leq 0$ for all age classes (but are not both identically zero) then clearly $\gamma'(0) > 0$. Thus in this case, which occurs very frequently in model equation for single species growth, only supercritical stable bifurcation can occur. Not all models of interest have these propeties however. Nor do all models of interest display supercritical stable bifurcation. Examples will be given below.

Further examples of the use of formula (7) can be found in Cushing (1984). The expansions (5) are also of use in studying the nonlinear effects of density on the equilibrium age distribution Cushing (1984).

Although formula (7) can be used to determine the direction of bifurcation (in the "generic" case when $\gamma'(0) \neq 0$, i.e. when the bifurcation is not locally vertical) it will be seen in the next subsection that there is often a simpler method of determining not only the direction of bifurcation, but other properties of the continuum C as well.

The stability results described here can also be extended to more general vital rates δ, β (Cushing (1985)).

2.3 The spectrum

In Section 2 above a very general existence result for positive equilibrium solutions of (1) with $F = nf$ was given, namely that there exists an unbounded continuum C of positive equilibrium pairs (n, u). In this section I want to return to the question raised in Section 2 and discuss the problem of describing the spectrum associated with C :

$$\sigma = \{n \in R | (n, u) \in C/(1, 0) \text{ for some } u \in B\}.$$

For each $n \in \sigma$, (1) has at least one positive equilibrium with $F = nf$. The closure $cl(\sigma)$ of σ is an interval $cl(\sigma) = [\sigma_i, \sigma_s]$, possibly infinite, which contains the critical value 1.

The problem is to find ways of describing $cl(\sigma)$ from a knowledge of the properties of the vital rates D and f as functions of density u. For each $n \in \sigma$, (1) with $F = nf$ has at least one positive equilibrium. Also of interest is the uniqueness or lack of uniqueness of this bounded equilibrium.

Since C is unbounded, either σ is unbounded or the set of positive equilibria from C

$\Sigma = \{u \in B | (n, u) \in C \text{ for some } n \in \sigma\}$

is unbounded (or both). A useful tool in studying these sets is a certain invariant on C which derives from the biological fact that in order to be at equilibrium a population's net reproductive rate must equal 1 , i.e. each individual (or unit) is ultimately replaced by exactly one individual offspring. Mathematically, this principle is derived by noting that equation (2a) is equivalent to $u(a) = u(0) \exp(-\int_0^a Dda)$ which when substituted into the birth equation (2b) yields, for nontrivial equilbria $u(0) \neq 0$, the invariant

$$1 = nN(u) \tag{8}$$

$N(u) := \int_0^A f(u) \exp(-\int_0^a Dds) \, da$.

In particular the invariant (8) holds for (n, u) from C and hence $\sigma_i \geq 0$. In general the endpoints σ_i, σ_s of the spectrum can lie anywhere within the range

$0 \leq \sigma_i \leq 1 \leq \sigma_s \leq +\infty$.

One can however often derive properties of N(u) from specified properties of the density dependent vital per unit rates D and f which permit one to compute σ_i and σ_s using (8). In fact formulas for these endpoints are rather easily derived in terms of N :

$\sigma_i = 1/N_s$, $\sigma_s = 1/N_i$ (extended sense)

where $N_i := \inf_\Sigma N(u)$ and $N_s := \sup_\Sigma N(u)$.

As a simple example consider the commonly occurring case when density effects cannot increase fertility nor decrease the death rate for all age classes:

$0 \leq f(u) \leq f(0) = \phi$, $0 \leq D(0) = \mu \leq D(u)$ for positive $u \in B$.

Then clearly $N(u) \leq 1$ by (3) so that $N_s \leq 1$ and hence $\sigma_i \geq 1$. It follows that $cl(\sigma) = [1, \sigma_s]$ and hence supercritical stable bifurcation occurs in this case.

Another frequently occuring case is when either the birth rate f drops to zero or the death rate D tends to infinity as the density u increases without bound. Then N(u) drops to zero as n increases without bound. Consequently if Σ is unbounded then $N_i = 0$. But either Σ or σ is unbounded and hence in either case $\sigma_s = +\infty$ or $cl(\sigma) = [\sigma_i, +\infty)$.

Models often have all of the properties above and hence the spectrum is precisely $\beta = (1, +\infty)$. In such cases it follows that a positive equilibrium exists for every inherent net reproductive rate n greater than 1 and at least those for n near 1 are stable.

The invariant (8) becomes particularly useful in the important case when the dependence of D and f on density u is through a dependence on a single positive linear functional $p = p(u)$ of density. Suppose $F = nf$ and $D = \mu + d$ where

$f = f(a, p(u))$, $d = d(a, p(u))$

$f: [0, A] \times R^+ \rightarrow R^+$, $d: [0, A] \times R^+ \rightarrow R^+$

and p is a strictly positive bounded linear functional:

$p: B \rightarrow R$, $p(u) > 0$ for $u \geq 0$, $u \not\equiv 0$.

Examples for p include total population size $p = \int_0^A u \, da$, weighted integrals $p = \int_0^A ku \, da$, $k \geq 0$, and point evaluations $p = u(a_0)$, $0 \leq a_0 < A$, all of which appear in the literature of density dependent population dynamics (Cushing (1985)) . Here p will be referred to as "population size" for simplicity.

In this case the invariant (8) becomes

$1 = nN(p(u))$

$N(x) := \int_0^A f(a, x) \, u_0(a) \, \exp(-\int_0^a d(s, x) \, ds) \, da$.

Thus the graph $P = \{(n, p(u)) \in R^2 | (n, u) \in C\}$ must lie on the planar graph of the relation $nN(x) = 1$. Hence a "bifurcation diagram" of the pairs (n, u) can often easily be drawn by either an explicit calculation of N or by a simple analysis of its graph. A couple of examples are given in the next subsection. More examples can be found in Cushing (1984).

Although not completely obvious it is easily shown that for equilibrium solutions $p(u_1) = p(u_2)$ if and only if $u_1 = u_2$ and that P is unbounded if and only if Σ is unbounded (Cushing (1985)) . Thus any uniqueness or boundedness properties derived from this graphical method are in fact valid for solutions of (2).

2.4 Some Examples

A standard procedure in studying bifurcation phenomena is to draw a "bifurcation diagram" in which some measure of the nontrivial solutions is plotted against the bifurcation parameter. For (2) a natural choice is to plot the B-norm

$|u|_\mu := \sup_{0 \leq a \leq A} |u/u_0|$ against n. The general results above concerning positive equilibria of (1) are represented in Figure 1.

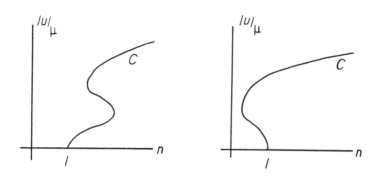

FIGURE 1. Supercritical Stable Bifurcation and Supercritical Unstable Bifurcation

For the case discussed at the end of the last subsection when F and D depend on a single positive linear functional p of density, it is more natural and convenient to construct a bifurcation diagram by plotting $p(u)$ against n as demonstrated in Figure 2.

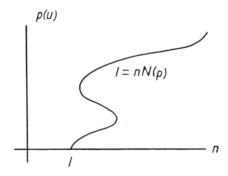

FIGURE 2. A bifurcation diagram in $(n, p(u))$ coordinates.

Two examples illustrating the two cases of supercritical stable bifurcation and subcritical unstable bifurcation will now be given.

Consider the "logistic" type case when the birth rate depends linearly on a functional p of density (Hoppenstaedt (1975)):

$$f = \phi(a)[1 - c(a)x]_+ \; , \; \int_0^A \phi \, u_0 \, da = 1 \; , \; c(a) > 0 \; , \; \phi(a) \geq 0$$

where $[x]_+ = x$ for $x \geq 0$ and $= 0$ for $x < 0$. Suppose that the death rate D increases with population size p :

$$d(a, 0) \equiv 0 \; , \; d_x(a, x) \geq 0 \; , \; 0 < a < A \; , \; x > 0 . \tag{9}$$

Then

$$N(x) = \int_0^A \phi[1 - cx]_+ \, u_0 \, \exp(-\int_0^a d(s, x) \, ds) \, da$$

has the following properties: $N(0) = 1$, $N(x) \equiv 0$ for $x \geq x_0 := \max 1/c$ and $N(x)$ is strictly decreasing on $0 < x < x_0$. Thus the graph of the relation $1 = nN(x)$ and hence the bifurcation diagram for this example appears as in Figure 3.

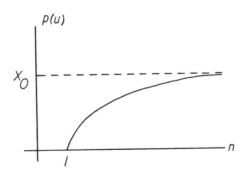

FIGURE 3 Bifurcation diagram for "logistic" model

It follows for this example that a positive equilibrium exists if and only if $n > 1$ in which case it is unique; that supercritical bifurcation occurs and hence the unique positive equilibrium is stable at least for n close to 1 ; and that $p(u) \to x_0$ as $n \to +\infty$.

As already remarked, stability does not in general persist globally along C. As will be seen below this fact can be illustrated using this particular example. A challenging mathematical problem is to study the global stability properties of examples such as this one ... global stability in both the sense of a global (nonlocal) stability analysis of a given density u for a specified fixed n and in the sense of a global analysis along C, i.e. for all $n \in \sigma$. Some results for special cases of this particular example have been obtained (see Marcati (1982) and Webb (1985)).

Supercritical stable bifurcation does not always occur, even in simple examples. Consider the so-called "depensation" model in which $F = b(e + x)/(c + x^2)$ for positive constants b, e, and c. The normalization (3) is accomplished by setting

$$f = \frac{c}{el} \frac{e + x}{c + x^2} , \quad n := bel/c , \quad l := \int_0^A u_0 \, da .$$

Suppose as a special case of (9) that the death rate is also "logistic-like" and linear in p: $d = x\psi(a)$, $\psi(a) > 0$. Then

$$N(x) = \frac{c}{el} \frac{e + x}{c + x^2} \int_0^A u_0 \, e^{-x\Psi(a)} \, da , \quad \Psi(a) := \int_0^A \Psi(\alpha) \, d\alpha .$$

Then $N(0) = 1$ and $N(+\infty) = 0$ so that $\sigma_s = +\infty$. If $N'(0) > 0$ then $N_s > 1$ and $\sigma_i < 1$ so that subcritical unstable bifurcation occurs. The condition $N'(0) > 0$ occurs for example if $1 > e\Psi(A)$.

Since $cl(\sigma) = [\sigma_i, +\infty)$, $0 < \sigma_i < 1$ in this case, there exist at least two positive equilibria for $\sigma_i < n < 1$. Roughly speaking one has a bifurcation diagram as given in Figure 4.

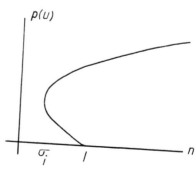

FIGURE 4 Bifurcation diagram for "depensation" model

An interesting problem for subcritical bifurcation cases such as this one is to determine what the asymptotic dynamics of (1) are. One might conjecture that in many circumstances the "upper" branch in this bifurcation diagram consists of stable positive equilibria while the lower branch consists of unstable positive equilibrium (from the local stability analysis above we know this is true near $(1, 0)$). If this is true the population is vulnerable to sudden collapses as n is decreased (below σ_i) and a hysteresis phenomenon is possible. Such phenomena have been suggested as being important in explaining some fishery collapses (see Clark (1976)).

The asymptotic dynamics of age-structured subcritically bifurcating population models such as this one pose challenging mathematical problems in that they can be quite complicated. I have for example investigated a simple depensatory type model in which preliminary investigations show that while it is true that the upper branch is stable and the lower branch is unstable as conjectured above, there also exist uncountably many positive periodic solutions each of which is approached by uncountably many trajectories.

These types of model equations are not without biological interest as they related to the so-called "Allee effect" when density effects on vital rates are not deleterious at low densities (although they might be at high densities) but in fact birth rates might increase with increased low level densities (e.g. due to an increased probability of mating) or death rates might decrease with increased low densities (e.g. due to the increased safety from herding or schooling). The periodic oscillations mentioned above in an age-structured subcritically bifurcating model suggest the possibility of a source of oscillations different from the familiar ones in population dynamics (time lags, predator-prey interactions, environmental periodicities, etc.) which is due to the internal age structure of the population together with an "Allee" or "depensatory" effect.

3 INTERACTING SPECIES

There is a rapidly growing literature on age-structed population interactions in which terms such as predation rates and preferences, competitive interaction coefficients, resource preferences and consumption rates, etc. are age specific (e.g. see Cushing and Seleem (1984, 1983); Diekmann, Nisbet, Gurney, and van den Bosch (1985); Gurtin and Levine (1979); Hastings (1983); Saleem (1984); and Webb (1985) just to cite a few). Such interactions can be modeled by coupling equations of the form (1) for the density of each species by assuming that the per unit birth and death rates depend on the densities of other species.

The approach taken here to the questions of equilibrium existence and stability for a single species as described above can be extended to the case of $m \geq 2$ interacting species in the following way. Suppose that a community of $m - 1 \geq 1$

interacting species with densities u_i , $1 \le i \le m - 1$, is known to have a positive equilibrium $u_i = u_i^0(a) > 0$, $0 \le a < A_i$. Under what conditions will the addition of an mth species result in a system with a positive equilibrium and when will this equilibrium be stable?

Let $s = s(a)$ be the density of the mth species and let $u = u(a)$ denote the vector of densities $(u_i) = (u_i)_{i=1}^{m-1}$. Then (u, s) denotes a vector of m densities. The equilibrium equations are

$$d_a u_i + u_i D_i = 0 , 0 < a < A_i$$

$$u_i(0) = \int_0^{A_i} u_i F_i \, da , \ u_i(A_i) = 0 \tag{10}$$

$$d_a s + s D_m = 0 , 0 < a < A_m$$

$$s(0) = \int_0^{A_m} s F_m \, da , \ s(A_m) = 0$$

where the vital rates $D_i = D_i(u, s)$, $F_i = F_i(u, s)$ depend on the densities u_i and s . Assume that the subcommunity of u_i governed by the "reduced" system

$$d_a u_i + u_i D(u, 0) = 0 , 0 < a < A_i$$

$$u_i(0) = \int_0^{A_i} u_i F_i(u, 0) \, da , \ u_i(A_i) = 0 \tag{11}$$

has a positive equilibrium u^0 in $B^{m-1} := B(\mu_1) \times ... \times B(\mu_{m-1})$, $\mu_i = D_i(0, 0)$. Here each μ_i satisfies the conditions on μ in the single species model equations above (but not necessarily with the same maximal age A_i) .

Assume further that the subcommunity equilibrium u^0 is nonsingular, i.e. that the linearization of (11) at u^0 possesses no nontrivial (i.e. nonidentically zero) solutions in B^{m-1} .

As in the approach for the single species case a parameter n is introduced into (10) by a normalization of the birth rate F_m of the species s and then equilibrium "pairs" $(n, (u, s))$ are sought which bifurcate from the trivial solution $(n, (u^0, 0))$. Specifically, let $F_m = n f_m$ where

$$\int_0^{A_m} s_0 f_m(u^0, 0)\, da = 1$$

$$s_0 := \exp(-\int_0^a D_m(u^0, 0)\, d\alpha) \, .$$

$$(12)$$

Then n has the biological interpretation of an inherent (i.e. low s density) net reproductive rate for species s when the other species u are at the subcommunity equilibrium u^0 .

The linear equations resulting from a linearization of (10) at $(u^0, 0)$ are not as easily studied as in the one species case. This is because the linearized equations for u_i has the form

$$d_a u_i + u_i D_i(u^0, 0) + u_i^0\, d_u D_i(u^0, 0)\,(u) = 0$$

in which there appears (unlike in the single species case linearized at $u = 0$) a "nonlocal" linear operator involving the Fréchet derivative of D_i with respect to u . Nonetheless an adequate linear theory for these and the remaining equations from the linearization of (10) can be established (including a requisite Fredholm theory) which permits the application of the global and local bifurcation theory techniques used in the single species case (see Cushing (to appear)).

The result is the following. Assume $A_i < \infty$ and that the operators

$$u_i D_i(u, s) \, , \quad \int_0^{A_i} u_i F_i(u, s)\, da$$

are $q \geq 1$ times continuously Fréchet differentiable at $(u, s) = (u^0, 0)$ as operators from B^m into $B(\mu_i)$ and R respectively. Let $F_m = n f_m$ where f_m satisfies the normalization (12). Then <u>the equilibrium equations</u> (10) <u>with</u> $F_m = n f_m$ <u>possess a continuum</u> C <u>of equilibrium "pairs"</u> $(n, (u, s))$ <u>with the following properties</u>

$$(1, (u^0, 0)) \in C$$

C <u>is unbounded in</u> $R \times B^m(\mu)$

<u>near</u> $(1, (u^0, 0))$ <u>the equilibrium pairs on</u> C <u>are positive and have the form</u>

$$n = 1 + \gamma(\varepsilon) \, , \quad u = u^0 + y(\varepsilon) \, , \quad s = \varepsilon s_0 + \varepsilon x(\varepsilon) \qquad (13)$$

<u>for</u> $\varepsilon > 0$ <u>small where</u> $\gamma(0) = 0$, $y(0) = 0$, $x(0) = 0$. <u>Thus near the bifurcation point the</u>

equilibria from C are positive.

The only significant difference between this multi-species result and the single species result above is that in general for the case of $m \geq 2$ interacting species the global, unbounded continuum C cannot be expected to consist entirely of positive equilibria, i.e. this continuum can leave the cone of positive functions in B^m. This fact can be seen for example from the classical Lotka-Volterra predator-prey or competition equations. If C does leave the positive cone, it must do so at a point

$(n', (u', s')) \neq (1, (u^0, 0))$ for which at least one u_i' identically vanishes or s' identically vanishes. This is because solutions of (10) are necessarily of one sign. Such a point corresponds to an equilibrium state for another "reduced" system or subcommunity of (10).

A linearized stability analysis completely analogous to that carried out for the single species case can be applied to (10). If the nonsingular equilbrium u^0 of the reduced system (11) is stable (as a solution of the corresponding dynamical equations for the reduced system), then the trivial equlibrium $(n, (u^0, 0))$ is stable if and only if $n < 1$ and the positive equilbria (13) are stable if and only if the bifurcation is supercritical (see Cushing (to appear)).

The direction of bifurcation can, as in the single species case, be determined from an explicit formula for $\gamma'(0)$:

$$\gamma(0) = \int_0^A m \ s_0 \ f_m^0 \int_0^a \nabla D_m^0 \ d\alpha da - \int_0^A m \ s_0 \ \nabla f_m^0 \ da$$

where the superscript "0" denotes evaluation at $(u, s) = (u^0, 0)$ and

$$\nabla D_m^0 := d_u \ D_m(u^0, 0)(d_\varepsilon \ y(0)) + d_s \ D_m(u^0, 0) \ (s_0)$$

(a similar formula holding for ∇f_m^0). If $\gamma'(0) > 0$ then the bifurcation is supercritical and stable.

The local stability near the bifurcation point in the case of supercritical bifurcation can be lost along C . Classical examples of predator-prey interactions between unstructured poulations which undergo Hopf bifurcations to stable limit cycles illustrate this fact.

Similar to the single species case properties of the continuum C can be deduced from the invariants

$$1 = nN_m(u, s) \ , \ 1 = N_i(u, s) \ , \ 1 \leq i \leq m - 1 \qquad (14)$$

which hold for any equilibrium $(u, s) \neq (0, 0)$ and in particular for the positive equilibria from C.

These are particularly useful when the vital rates D_i, F_i all depend on the densities (u, s) through strictly positive linear functionals (or "population sizes") $p_i: B(\mu_i) \to R$ as follows:

$$D_i = \mu_i + d_i(a, p(u), p_m(s)) \; , \; F_i = F_i(a, p(u), p_m(s))$$

$$p(u) := (p_i(u_i))_{i=1}^{m-1} \; , \; d_i(a, 0, 0) \equiv 0$$

where $d_i, F_i: [0, A_i] \times R^m \to R$. Then (14) implies that the graph $\{(n, (p(u), p_m(s))|(n, (u, s)) \in C\}$ lies on the graph of the relations

$$1 = nN_m(x, x_m) \; , \; 1 = N_i(x, x_m) \; , \; 1 \leq i \leq m - 1 \tag{15}$$

where $x = (x_i)_{i=1}^{m-1}$ and

$$N_i := \int_0^{A_i} F_i(a, x, x_m) \exp(-\int_0^a (\mu_i + d_i(\alpha, x, x_m)) \, d\alpha) da$$

$$N_m := \int_0^{A_m} f_m(a, x, x_m) \exp(-\int_0^a (\mu_m + d_m(\alpha, x, x_m)) \, d\alpha) da$$

Let $p^0 := (p_i(u_i^0))_{i=1}^{m-1}$. By the normalization (12) and from the invariants for the equilibrium u^0 of the reduced system one has

$1 = N_i^0 \; , \; 1 \leq i \leq m$.

The relations (15) can be used, amongst other things, to determine the direction of bifurcation. Suppose that the last $m-1$ equations in (15) are solved for $x = x(x_m)$, $x(0) = p^0$, and that these solutions are substituted into the first equation. Supercritical and hence stable bifurcation occurs if

$$d_{x_m} N_m(x(x_m), x_m)|_{x_m=0} < 0 .$$

Consequently if the Jacobian

$$J := (d_i N_j^0)_{1 \leq i, j \leq m-1} \; , \; d_i := d_{x_i}$$

is nonsingular, then supercritical bifurcation occurs if

$$((d_i N_m^0), d_m N_m^0) \cdot (-J^{-1}(d_m N_i^0), 1) < 0$$

where the dot "·" denotes the usual vector inner product.

These existence and stability conditions yield constraints on the density dependence of the net reproductive rates N_i of all species under which the species s can "invade" and coexist (at low densities) with the stable subcommunity fo species u_i in the sense that there exists a stable positive equilibrium of the full m species community. This interpretation is valid at least near bifurcation (i.e. for low s densities).

As an example consider the often studied case of $m = 2$ interacting species. The conditions above for supercritical bifurcation reduce to

$$d_1 N_1^0 \neq 0 , d_2 N_2^0 - d_1 N_2^0 d_2 N_1^0 / d_1 N_1^0 < 0 . \tag{16}$$

Two species interaction models for unstructured populations are usually categorized by the interspecies effects that total population size P_i has on the per capita growth rates P_i'/P_i . For example species i is said to prey on species j if $d_{P_j} (P_i'/P_i) > 0$ and $d_{P_i}(P_j'/P_j) < 0$ or these two species are said to compete if the first inequality is reversed.

For age-structured models appropriate assumptions need to be made on the dependencies of the age-specific death and birth rates of each species on the density of the other (or in the case being considered here on the population sizes $p_i(u_i)$ of the other species) according to the type of interaction being modeled. The conditions (16) suggest, on the other hand, that for age-structured populations an appropriate categorization of the type of interaction might instead be made according to each species' effect on the net reproductive rate of the other.

One might for example distinguish the two fundamental cases

$$d_1 N_2 > 0 , d_2 N_1 < 0 \qquad \text{(Predator-prey)} \tag{17}$$

$$d_1 N_2 < 0 , d_2 N_1 < 0 \qquad \text{(Competition).} \tag{18}$$

If the common intraspecific density effect assumptions $d_i N_i < 0$ are made whereby a species' net reproductive rate at equilibrium decreases with increases in population size, then the supercritical stable bifurcation condition (16) is equilvalent to the determinant-like condition

$$d_1 N_2^0 d_2 N_1^0 - d_1 N_1^0 d_2 N_2^0 < 0 . \tag{19}$$

For the predator-prey case (17) this condition is always met. Thus, a predator s can always survive on a prey u_1 (at least at low predator densities) if its inherent net reproductive rate n when the prey u_1 is at stable equilibrium is larger than one.

For the competition case (18) however stable coexistence is possible only if (19) holds. Condition (19) is very analogous to the classical condition for two species coexistence in unstructured populations which requires that the determinant of the so-called "community matrix" be positive. It is susceptible to the same interpretation as that for the classical unstructured population growth equations, namely that stable coexistence requires intraspecific competition (as measured by the product $d_1 N_1^0 d_2 N_2^0$) to be stronger than interspecific competition (as measured by the product $d_1 N_2^0 d_2 N_1^0$).

4. PERIODIC OSCILLATIONS
4.1. The Autonomous Case.

I have until now considered only the primary bifurcation in the bifurcation scenario described in the Introduction, namely the bifurcation of equilibrium states from zero densities as a function of the inherent net reproductive rate n . But I have repeatedly pointed out that even when this primary bifurcation is supercritical and hence stable and even when the existence of positive equilibria is global it is not necessarily true that stability persists globally. For example it can happen when when n reaches a second critical value $n_0 > 1$ that the stable equilibria destabilize and a Hopf-type bifurcation to nonsteady, time periodic densities $u = u(t, a) > 0$ occurs.

This problem was studied for the general age-structured equations (1) in (see Cushing (1983)) (the results there can be extended in a straightforward way to systems (10)). Rather than describe the results in Cushing (1983) here, I will simply illustrate them by an example based on the first example in Section 4 above.

Consider the following case of Hoppensteadt's model studied in Section 4 above:

$$d_t u + d_a u + uD = 0 , \ 0 < a < A = +\infty$$
$$u(t, 0) = n \int_0^A \phi(a)[1 - p(u)]_+ u(t, a) \, da \tag{20}$$

$$p(u) := \int_0^A \phi(a) u(t, a) \, da$$

$D = \text{constant} > 0$, $\phi(a) \geq 0$, $\int_0^A \phi(a) \, e^{-Da} \, da = 1$.

The equilibrium solutions can easily be found by direct calculation:

$$u(a) = (n - 1)e^{-Da}/n \tag{21}$$

and the continuum C consists of these equilibria for $n \geq 1$. Clearly the bifurcation is supercritical and hence stable. Positive equilibria exist and are unique for all $n > 1$.

It was shown in Cushing (1983) that if the "characteristic equation" of the linearization of (20) at the equilibrium (21) possesses a conjugate pair of complex roots which transversally crosses the imaginary axis (at a nonzero point) as n increases through a critical value n_0 then bifurcating from the equilibria (21) will be a continuum of nontrivial time periodic solutions of (20) in the usual Hopf-like manner. The characteristic equation for the example being considered here is

$$C(z) := 1 - (2 - n) \int_0^A \phi(a) \, e^{-Da} \, e^{-za} \, da = 0 .$$

Depending on the properties of the age specific birth rate coefficient ϕ , this equation can have a conjugate pair of complex roots which transversally cross the imaginary axis at some critical value of n . This is perhaps best illustrated here by a concrete example.

The function $\phi(a)$ in (20) describes the inherent age-specific birth rate of the species. That $p(u)$ is as defined in (20) means that the adverse effects of population density on fertility are in this model age-specific and caused by the more fertile age classes themselves. A specific type of kernei ϕ which is frequently used in qualitative studies is

$$\phi(a) = \frac{1}{k!} (\frac{k}{M} + D)^{k+1} a^k e^{-ka/M} , M > 0 , k = 1,2,3,\dots . \tag{22}$$

Such a kernel implies that fertility monotonically increases from zero (at age zero) with age until peaking at the age M of maximum fertility before monotonically decreasing to zero as age continues to increase without bound. The peak is sharper, i.e. maximum fertility is narrowly distributed around age M , for larger k . The constants appearing in ϕ are chosen so that the normalization in (20) holds.

For such a kernel the characteristic equation reduces to

$$1 - (2 - n)(q/(q + z))^{k+1} = 0 , q := \frac{k}{M} + D$$

whose roots are $z = q(-1 + (2 - n)^{1/(k+1)})$.

For $1 < n < 3$ all roots lie in the left half plane for all k. For $k = 1$, all roots lie in the left half plane for all $n > 1$. In these cases the equilibrium (20) is stable.

However, if the age-specific fertility function ϕ is sufficiently narrowly peaked at $a = M$, i.e. if $k \geq 2$, then a pair of conjugate roots z transversally crosses into the right half plane at the critical value

$$n_0 := 2 + \sec^{k+1}(\pi/(k + 1)) > 3$$

where

$$z = i\theta , \ \theta = q(n_0 - 2)1/(k+1)\sin(\pi/(k + 1)) .$$

Thus, by the results in Cushing (1983), a Hopf bifurcation from the equilibria (21) to nontrivial time periodic solutions of (20) occurs as n increases through n_0 for the kernels (22) with $k \geq 2$.

The bifurcation results in Cushing (1983) for equations of the type (1) are purely existence results. A stability analysis awaits future research. Perhaps the semi-group theory developed in Webb (1985) would be useful for this purpose.

4.2 The Nonautonomous Case.

The model equations studied so far have been all time autonomous, i.e. the equations do not depend explicitly on time t. Even though it is widely recognized (even by such early founding fathers of mathematical population dynamics as Volterra, Kostitzin and others) that important modeling parameters can and often do fluctuate significantly in time, there has been relatively speaking little research done on nonautonomous model equations. This is undoubtedly due to a large extent to the difficulties in mathematically analysing nonautonomous equations as compared to autonomous equations.

There are roughly two broad classes of nonautonomous equations which are being studied in an increasing amount of literature: stochastic equations and time periodic equations. With regard to the age-structured model equations (1) and (10) however there seems to have been very little research done on the difficult problems arising when the vital rates D, F are allowed to fluctuate stochastically. See however Witten (1983) for a study of the linear case.

If the vital rates R and D are allowed to oscillate periodically with a common period, a natural question which arises from the approach taken here is whether the bifurcation results for positive equilibria in the autonomous case can be extended to positive time periodic densities in the periodic case. For the unstructured population case this has been successfully carried out for general systems of interacting populations in Cushing (1982) . For the age-structured case of a single species governed by (1) I have (in a forthcoming paper (Cushing (to appear); also see Cushing (1986)) extended the local bifurcation results described above for the autonomous case to the periodic case. This extension involves much more sophisticated mathematics than needed for the autonomous case (such as the use of results and methods concerning power compact, strictly positive operators on Banach lattices in order to handle the necessary linear equations, which are in this case periodic partial differential equations). A global analysis for periodic versions of (1) and an analysis of periodic versions of systems (10) remains to be carried out.

5 CONCLUDING REMARKS

There are of course many possibilities for further analysis of age-structured population growth models using the approach taken in this paper. For example there is the problem of studying the dynamics of the model equations (1) or (10) when the growth rate v of the physical parameter x depends on t and, as would certainly be the case in many biologically reasonable cases, on the densities u_i . This assumption introduces further nonlinearities into the equations which pose new mathematical difficulties because the characteristics of the resulting hyperbolic equations then depend on the solutions.

Another example of an important and interesting problem is to introduce spatial heterogeneity and allow for spatial diffusion by the populations as well as for age structure within the population. Spatial diffusion by unstructured populations has been extensively studied in recent years.

There are of course also many interesting, challenging, and important mathematical problems involved in modeling and analysing population growth dynamics for specific types of populations and for specific types of multi-species interactions. This is especially true for age-structured populations or for populations structured by other internal variables.

In this paper however I have not discussed in detail any particular type of population or community of populations, but have instead emphasized some very general results which apply to very general dynamical growth models. I have done this with the hope that the approach taken here, which utilizes bifurcation theory and techniques, has not only suggested a way to apply the powerful analytical techniques available from this theory, but has shown how this theory can provide a unification of some fundamental properties shared in common by a very broad class of population dynamical models.

REFERENCES

Clark, C.W., 1976. Mathematical Bioeconomics, Wiley, New York.
Cushing, J.M., 1980. Model stability and maturation periods in age structured
  populations, J. Theo. Biol. 86:709-730.
_____, 1981. Volterra Integrodifferential Equations in Population Dynamics, in
  Mathematics in Biology (M. Iannelli ed.), Liguori Editore, Naples.
_____, 1982. Periodic Kolmogorov systems, SIAM J. Math. Anal. 13, no. 5, 811-827.
_____, 1983. Bifurcation of time periodic solutions of the McKendrick equations with
  applications to population dynamics, Comp. & Math. Appl. No. 3, 9: 459-478.
_____, 1985. Equlibria in structured populations, J. Math. Biol. 23: 15-39.
_____, Existence and stability of equilibria in age-structured population dynamics,
  1984. J. Math. Biol. 20: 259-276.
_____, Periodic McKendrick equations for age-structured population growth, 1986.
  Comp. & Math. Appl. 415: 513-526.
_____, Periodic McKendrick equations, to appear in Comp. & Math. Appl.
_____, Equilibria in systems of interacting structured populations, to appear J. Math.
  Biol.
_____ and M. Saleem, A competition model with age structure, 1984. In Mathematical
  Ecology (S.A. Levin and T.G. Hallam, ed.), Lec. Notes in Biomath. 54: 178-192.
_____, Competition and age-structure, 1983. In Population Biology (H.I. Freedman &
  C. Strobeck ed.), Lec. Notes in Biomath. 52: 210-217.
Diekmann, O., Nisbet, R.M., Gurney, W.S.C., and van den Bosch, F. 1985. Simple
  mathematical models of cannibalism: a critique and a new approach, Centre for
  Math. & Comp. Sci., Report no. Am-R8505, Amsterdam.
Gurtin M. and MacCamy, R.C., 1974. Nonlinear age dependent population dynamics,
  Arch. Rat. Mech. Anal. 54: 281-300.
Gurtin, M., The Mathematical Theory of Age-Structured Populations, to appear.
Gurtin, M. and Levine, D.S., 1979. On predator-prey interactions with predation
  dependent on age of prey, Math. Biosci. 47: 207-219.
Hastings, A. 1983. Age dependent predation is not a simple process I: continuous time
  models, Theo. Pop. Biol. 23: 347-362.
Hoppensteadt, F. 1975. Mathematical Theories of Populations: Demographics,
  Genetics, and Epidemics, SIAM Conf. Series on Appl. Math., Philadelphia.
Marcati, P. On the global stability of the logistic age dependent population
  equation, 1982. J. Math. Biol. 15: 215-226.
McKendrick, A.G. 1926. Applications of mathematics to medical problems, Proc. Edin.
  Math. Soc. 44: 98-130.
Nisbet, R.M. and Gurney, W.S.C. 1982. Modelling Fluctuating Populations, Wiley, New
  York.
Oster, G., 1977. Lectures in Population Dynamics. In Lectures in Applied Mathematics,
  AMS, 16: 149-170.
Saleem, M. 1984. Predator-prey relationships: Indiscriminate predation, J. Math. Biol.
  No. 1, 21: 25-34.
Sinko, J.W. and Streifer, W. 1967. A new model for the age-size structure of a
  population, Ecol. 48: 910-918.
von Forester, H. 1959. Some remarks on changing populations. In The Kinetics of
  Cellular Proliferation (F. Stholman Jr. ed.), 382-407, Grune & Stratton, New York,
Webb, G.F. 1985. Theory of Nonlinear Age-Dependent Population Dynamics,.
  Monographs in Pure and Applied Mathematics Series, vol. 89, Marcel Dekker,
  New York.
_____, Logistic models of structured population growth,1986. Comp. & Math. Appl.
  12A, Nos. 4/5: 527-539.
Witten, Matthew. 1983. On stochasticity in the von Forester hyperbolic partial
  differential equation system. Further applications to the modeling of an
  asynchronously dividing cellular system, Comp. & Math. Appl. No.3, 9: 447-458.

PART IV - COMMUNITY ECOLOGY

YOUNG PREDATION AND TIME DELAYS

M. SALEEM, S.U. SIDDIQUI and V. GUPTA
Department of Mathematics, HBTI, Kanpur, 208016, INDIA

1 INTRODUCTION

For nearly two decades, age specific predation has been a subject of interest to many researchers. Incorporating the fact that predators do not eat all ages nor all sizes of prey indiscriminately (various examples ranging from molluscs to insects to fish illustrate this point), May (1974) and Smith and Mead (1974) proposed some of the early models for predator-prey interactions in which the predator species ate either only the young (or eggs) or only the adults of prey. An example where predators eat only the eggs (or young) is that of Walleyes and yellow perch discussed by Nielsen (1980) while that for the latter case is of periodical cicada studied by Lloyd and Dybas (1966).

For the simple Lotka-Volterra type interaction terms, predation on only juveniles or only on adults has been shown to act as a stabilising factor. In fact, the model which is originally neutrally stable becomes stable for all values of parameters (see May (1974); Smith and Mead (1974)). Another model suggested by Gurtin and Levine (1979) reached an opposite conclusion indicating that age-dependent predation may be destabilising. Since then numerous authors have considered models, some suggesting that age-dependent predation is stabilising while others suggest the opposite destabilising. We refer to Coleman and Frauenthal (1983); Frauenthal (1983); Gurtin and Levine (1979); Hastings (1983-84); Levine (1981); Saleem (1983-84, 1986); Thompson et al. (1982) for the relevant literature on this subject.

Time-delays in the growth dynamics of populations have been considered by many investigators (see Caperon (1969); Caswell (1972); Cushing (1977, 1982a, b); Freedman and Rao (1983); Gopalsamy (1984a; 1984b); May (1973, 1974); May, Conway, Hassell, and Southwood (1974); McDonald (1976); Nicholson (1957); Shukla (1983); Smith (1974); Solimano and Beretta (1983); Volterra (1927, 1931)). In almost all these papers the interactions among the species have been of classical Lotka-Volterra types although the stability analyses of the equilibria have been of local as well as of global nature. Hastings (1983) introduced time-delays in the growth dynamics of prey through a juvenile period and studied the interplay between time-delays (in the prey dynamics) and the predator's functional response. Based on the analysis of his different models and having support from some earlier findings (Cushing (1977); Cushing and Saleem (1982); May (1974)), he argued that time-delays were not necessarily destabilising, that age-specific predation was a complicated matter, and the results about stability and limit cycles, etc. were very model-dependent. Hastings (1984) also considered the predator-prey models with both species having delays in the recruitment.

There seems to be very little work done in young predation models along the lines where delays have been introduced in the growth rate response of predators. In two recent papers (Saleem 1983-84), we studied these cases by first introducing maturation periods in the fecundity of predator and, second, by considering the effects of all previous prey-egg-levels on predator's birth rate (per unit individual per unit time).

In the present paper, we formulate and analyze a mathematical model which considers both aspects, namely (i) the age-structure in both species and (ii) the effects of all previous prey-egg-levels on the birth rate of predators. The former aspect allows introduction of delays into the fecundity of predators. The main concern of the current paper is an answer to the following question.

"How would delays in predator's growth rate response affect the young predation stability results?"

To answer this question, we consider three special cases which lead to three different model-systems whose results either have been reported or will be reported elsewhere. In the present paper we will make a comparative study of these results. Throughout this paper the term stability will mean (local) asymptotic stability.

The model is defined in section 2. Section 3 contains some special cases of the model while in section 4 the stability results are stated. Conclusions may be found in section 5.

2 MODEL EQUATIONS

We assume prey and predator populations with age structure specified by age-distributions $p_1(a, t)$ and $p_2(a, t)$ respectively of individuals of age a at time t. The total population sizes $P(t)$ and $Q(t)$ of prey and predator at time t respectively may be found by an integration over all ages as follows:

$$P(t) = \int_0^\infty p_1(a, t)\, da \quad \text{and} \quad Q(t) = \int_0^\infty p_2(a, t)\, da \ .$$

We assume the processes of aging and death of individuals already in the two populations at time t to be described by the balance equations (known as McKendrick Von Foerester equations)

$$\frac{\partial p_1}{\partial t} + \frac{\partial p_1}{\partial a} = -\mu p_1$$

$$\frac{\partial p_2}{\partial t} + \frac{\partial p_2}{\partial a} = -\eta p_2$$

$$a > 0\, , \ -\infty < t < \infty \qquad (2.1)$$

where μ and η are positive constants and denote the death rates (per unit individual) of prey and predator respectively.

The recruitments to prey and predator populations are assumed to be given by the additional renewal equations respectively.

$$B_1(t) = \int_0^\infty f_1(a, t)\, p_1(a, t)\, da$$

$$(2.2)$$

$$B_2(t) = \int_0^\infty f_2(a, t)\, p_2(a, t)\, da$$

where f_1 and f_2 are the fecundity rates (per unit individual) of prey and predator species respectively. More specifically, f_i is the expected rate at which the off-spring are born to an individual of i-th species of age a at time t.

An integration of (2.1) yields

$$p_1(a, t) = p_1(0, t - a)\, e^{-\mu a}$$

$$(2.3)$$

$$p_2(a, t) = p_2(0, t - a)\, e^{-\eta a} \quad .$$

Furthermore, an integration of (2.1) from $a = 0$ to $a = +\infty$ gives

$$p_1(0, t) = \frac{dP}{dt} + \mu P$$

$$(2.4)$$

$$p_2(0, t) = \frac{dQ}{dt} + \eta Q$$

Substituting the expressions for $p_i(0, t - a)$ from (2.4) into (2.3) and then after replacing $p_i(a, t)$ in (2.2) by (2.3) followed by an integration by parts, we obtain the following expressions for $B_i(t)$ (see Cushing and Saleem (1982); Saleem (1983)).

$$B_1(t) = \int_0^\infty \frac{\partial f_1}{\partial a}\, P(t - a)\, e^{-\mu a}\, da$$

$$(2.5)$$

$$B_2(t) = \int_0^\infty \frac{\partial f_2}{\partial a}\, Q(t - a)\, e^{-\eta a}\, da$$

Considering the fact that predators eat only the eggs (or young) of prey, we assume (as in Levine (1981); Thompson et al. (1982)), the following relations involving $p_1(0, t)$ and $B_i(t)$.

$$p_1(0, t) = \frac{B_1(t)}{1 + k\,Q(t)}$$

$$p_2(0, t) = B_2(t)$$

(2.6)

Making use of (2.4), (2.5) and (2.6), we obtain the following system of integrodifferential equations

$$\frac{dP}{dt} = -\mu P + \frac{1}{1 + kQ} \int_0^\infty \frac{\partial f_1}{\partial a} P(t - a)\, e^{-\mu a}\, da$$

$$\frac{dQ}{dt} = -\eta Q + \int_0^\infty \frac{\partial f_2}{\partial a} Q(t - a)\, e^{-\eta a}\, da$$

(2.7)

for prey and predator species respectively. The fact that the predator fecundity rate f_2 depends on the prey-egg-levels makes (2.7) a coupled system.

The fecundity rate f_2 of predators will be assumed to be a function of t through its dependence on another function $E(t)$ (see (2.9) below for expression) which describes the accumulated effects of all previous prey-egg-levels $p_1(0, s)$ $(s \le t)$ on the current fecundity rate f_2 at time t.

We assume

$$f_2(a, t) = c\, w_2(a)\, E(t)$$

(2.8)

where c is a positive constant representing the birth modulus of predator. The function w_2, known as maturation function for predators and describing the effects of age a on the fecundity f_2, is assumed to be continuously differentiable for $a \ge 0$. Additional assumptions will be imposed on w_2 later. The function $E(t)$ is defined as

$$E(t) = \int_{-\infty}^t K(t - s)\, p_1(0, s)\, ds$$

(2.9)

where the kernel K is the weight function which represents the weighted (or distributed) effects of all previous prey egg-levels $p_1(0, s)$ $(s \le t)$ on the current fecundity rate f_2 at time t.

We assume the function K is non-negative, piecewise continuous, satisfying

$$K \in L^1_+$$

and normalised such that

$$|K|_1 = \int_0^\infty K(S) \, ds = 1 .$$

Substituting for f_2 from (2.8) into (2.7) and making use of (2.5) and (2.6) we get the following system of integrodifferential equations

$$\frac{dP}{dt} = -\mu P + \frac{1}{1 + kQ} \int_0^\infty \frac{\partial f_1}{\partial a} P(t - a) \, e^{-\mu a} \, dt ,$$

(2.10)

$$\frac{dQ}{dt} = -\eta Q + c \int_0^\infty \frac{K(S)}{1 + kQ(t - s)} \int_0^\infty \frac{\partial f_1}{\partial a} P(t - s - a) \, e^{-\mu a} \, da \, ds \int_0^\infty w_2'(a) \, Q(t - a) \, e^{-\eta a} \, da$$

to be studied in this paper.

3 SOME SPECIAL CASES

We will consider, in this section, three special cases of (2.10) in terms of three different model-systems A, B, and C by assuming some specialised forms for functions w_2 and K.

MODEL A [A case with no age structure in the predator population and no previous egg-level-effects] .

Let us consider $w_2(a) \equiv 1$ and $K(s) = \delta_0$, the Dirac delta function at $s = 0$. The assumption $w_2(a) \equiv 1$ amounts to considering that no age structure is present in the predator population. Assuming $K(s) = \delta_0(s)$, implies that no effects of previous prey-egg-levels on the current fecundity rate of predator are considered. These assumptions on w_2 and K reduce (2.10) to the following system

$$\frac{dP}{dt} = -\mu P + \frac{1}{1 + kQ} \int_0^\infty \frac{\partial f_1}{\partial a} P(t - a) \, e^{-\mu a} \, da ,$$

(3.1)

$$\frac{dQ}{dt} = -\eta Q + \frac{cQ}{1 + kQ} \int_0^\infty \frac{\partial f_1}{\partial a} P(t - a) \, e^{-\mu a} \, da .$$

We refer to (Saleem (1983)) for the stability analysis of the non-negative equilibria of (3.1).

MODEL B [A case with age structure in the predator population but no previous egg-levels effects] .

Let us assume $K(s) = \delta_0(s)$, the Dirac delta function at $s = 0$, in order to neglect the previous egg-level-effects. This assumption converts (2.10) to the system

$$\frac{dP}{dt} = -\mu P + \frac{1}{1+kQ}\int_0^\infty \frac{\partial f_1}{\partial a} P(t-a)\, e^{-\mu a}\, da ,$$

$$\frac{dQ}{dt} = -\eta Q + \frac{c}{1+kQ}\int_0^\infty \frac{\partial f_1}{\partial a} P(t-a)\, e^{-\mu a}\, da \int_2^\infty w'(a)\, Q(t-a)\, e^{-\eta a}\, da .$$

$$(3.2)$$

The stability of non-negative equilibria of (3.2) is studied in detail in Saleem (1984).

MODEL C [A case with previous egg-level-effects but no age structure in the predator population]

In order to consider a situation where no age-structure is assumed in the predator population, we let $w_2(a) \equiv 1$. With this choice of the function w_2 , the system (2.10) changes to

$$\frac{dP}{dt} = -\mu P + \frac{1}{1+kQ}\int_0^\infty \frac{\partial f_1}{\partial a} P(t-a)\, e^{-\mu a}\, da$$

$$\frac{dQ}{dt} = -\eta Q + cQ\int_0^\infty \frac{K(s)}{1+kQ(t-s)}\int_0^\infty \frac{\partial f_1}{\partial a} P(t-a-s)\, e^{-\mu a}\, da\, ds$$

$$(3.3)$$

a system studied in detail in Saleem (1986).

For highlighting certain conclusions of this paper, we will mention some of the stability results for model-systems A, B, and C in the next section.

4 RESULTS

We assume the fecundity rate function f_1 of prey to be of the form

$$f_1 = f_1(a) = bw_1(a) \tag{4.1}$$

where b , a positive constant, is the birth modulus of prey. The function, w_1 , describing the effects of age a on the prey fecundity, is assumed to be continuously differentiable, bounded for $a \geq 0$, and to satisfy

$$w_1(+\infty) = w_1(0) = 0 \ , \ w_1(a) \geq 0 \ \text{ and } \ \int_0^\infty a \, w_1(a) \, e^{-\mu a} \, da < +\infty \qquad (4.2)$$

The net reproductive rate R of prey, which gives the expected number of offspring born to an individual in its life-time, is given by

$$R = \int_0^\infty f_1 \, e^{-\mu a} \, da = b w_1^*(\mu)$$

where $*$ denotes the Laplace-transform.

Models A, B, and C have the same non-negative equilibria:

$$e_0 = (P, Q) = (0, 0) \ \text{ and } \ e_+ = (P, Q) = (\frac{1}{c\mu \, w_2^*(\eta)}, \ \frac{R-1}{k})$$

where $*$ denotes the Laplace-transform. Note that $w_2^*(\eta) = \frac{1}{\eta}$ when $w_2(a) \equiv 1$. We now mention the stability results for e_0 and e_+.

(i) Instability Result for e_0 of models A, B, and C.

Theorem 0: The trivial equilibrium e_0 is unstable provided $R > 1$ but $\simeq 1$. (If the net reproductive rate R of prey is greater than but close to 1 then both species will not go to extinction).

(ii) Stability Results for e_+ of Model A

Assume (4.2) and $R > 1$.

Theorem 1: The positive equilibrium e_+ is (locally) asymptotically stable. (If the net reproductive rate R of prey is greater than 1 then both species will stably coexist).

MODEL B: Assume $w_2(a) = \frac{1}{m_2^2} \, e^{-a/m_2}$, $a \geq 0$ where m_2, a positive constant, is the age at which the fecundity of the predator maximizes. We will call $a = m_2$ the maturation period of the predator (for details see (Cushing and Saleem (1982)).

Theorem 2: The positive equilibrium e_+ is (locally) asymptotically stable provided $m_2 > \frac{2}{\eta}$. (Both species will stably coexist if the maturation period of the predator is greater than a certain number $\frac{2}{\eta}$).

MODEL C: Assume $w_1(a) = \dfrac{1}{\mu}(\mu + \dfrac{1}{m_1})^2 \, a \, e^{-a/m_1}$ and $K(a) = \dfrac{1}{T^2} a \, e^{-a/T}$, $a \geq 0$, such that $\mu \, w_1^*(\mu) = 1$ and $|K|_1 = 1$. The latter two conditions on w_1 and K are for

mathematical reasons. The function w_1 has almost all the features that one would like to have in a fecundity function in the sense that it ensures that no individual of small age as well as that of large age reproduces and also that at a certain age, $a = m_1$, an individual reproduces at a maximum. The second function K represents a case when the maximum influence on the birth rate of predators at any time t is due to the prey egg-level at time $t - T$. (Note that the maximum of K occurs at $t = T$) . For use in the following, we call T , the crucial-egg-consumption period for predator.

Theorem 3: (a) Let $\gamma = \mu + \dfrac{1}{m_1}$. If

$$\frac{\gamma R}{2(R-1)} < \eta < \frac{4\gamma R}{5(R-1)} \tag{4.3}$$

then the positive equilibrium e_+ is (locally) asymptotically stable provided

$$\frac{\eta(R-1)}{2\gamma^2 R} < T < \frac{2}{5}\frac{1}{\gamma} . \tag{4.4}$$

(If the death rate η and the crucial-egg-consumption period T of predators lie in intervals given by (4.3) and (4.4) respectively then both species will stably coexist).

(b) For $T \simeq 0$, the positive equilibrium e_+ is (locally) asymptotically stable (unstable) if

$$\eta < \frac{3\gamma R}{(R-1)} \; (> \frac{3\gamma R}{(R-1)}) \tag{4.5}$$

(For small crucial-egg-consumption-period T , both species will stably exist (not stably exist) if the death rate η of predators is less than (greater than) a fixed number given by (4.5)) .

(c) For $T \simeq \infty$, the positive equilibrium e_+ is unstable. (For large crucial-egg-consumption-period T , both species will not stably exist). The pictorial representation of the results of Models A, B, and C is given below in Figures (a), (b) and (c) respectively. S stands for stable while U stands for unstable.

A comparative study of these results is made in the following section.

5 CONCLUSIONS

We have formulated a general model-system (2.10) representing the predator-prey interactions with both species age structured and predators eating only the eggs (or young) of the prey. It is further assumed that the predator fecundity rate at any time t is affected not only by the prey egg level, $p_1(0, t)$, at time t but also on all previous prey egg levels, $p_2(0, s)$, $(s \leq t)$. The weighted (or distributed) effects of these earlier egg levels is represented by a weight function K.

Three special cases giving rise to three different models of the main model system (2.10) are considered. A comparison of the results of these models reflects on the basic question of this paper i.e., "How would delays in predator growth rate response affect the young predation stability results?"

It is seen that the net reproductive rate R of prey must be greater than but close to 1 in order to avoid the extinction of both species (predator and prey) [Theorem 0].

In Theorem 1 it is shown that if no age-structure is assumed in the predator population then both species will always stably coexist without any conditions on system parameters. (See Figure A).

FIGURE A. System Parameters

It is demonstrated next [Theorem 2] that in the case of an age-dependent fecundity rate fluctuation of predator, the two species will stably co-exist if the maturation period m_2 of predators is greater than a fixed number $\frac{2}{\eta}$. When the maturation period m_2 is less than the same number $\frac{2}{\eta}$, the two species may or may not exist stably (see also Figure B).

FIGURE B. Predator's Maturation Period m

Thus we note that introduction of age-structure in mathematical models may complicate the issue with respect to model-stability. Furthermore, we have a mathematical model (3.2) whose result is in contrast to the commonly held tenet in ecological literature that large delays cause instabilities. Indeed, we have a situation which suggests just the opposite i.e. large delays cause stability.

Comparing the results of Theorems 1 and 3 (a) we find that certain conditions on the death rate η and the crucial-egg-consumption period T (see (4.3) and (4.4)) are required for stable coexistence of both species provided the effects of previous prey egg-levels on the current birth rate of predators are considered [Theorem 3(a)]. Contrary to this, Theorem 1 says that no such conditions for coexistence are required if previous egg level effects are not considered. We therefore see that introduction of time-delay into the growth rate response of predators may worsen the situation.

Unlike Theorem 1, we again have a condition (4.5) on the death rate η of predators, which needs to be satisfied in order to ensure the coexistence of both species when the crucial- egg-consumption period T is small enough [Theorem 3(b)].

Theorem 3(c) states that for a large crucial-egg-consumption period T the two species will not stably exist, hence the possibility that both species may suffer oscillations exists.

Interestingly enough from Figure (C) we notice the following:

(i) Keeping the death rate η of predators fixed somewhere in the interval (4.3), the delays T must lie in a bounded interval (4.4) for the stable coexistence of both species.

(ii) Fixing the death rate h small (close to zero) the positive equilibrium changes from stability to instability for increasing delays T as T crosses the critical value

$$T_c = \frac{3m_1}{4(\mu\, m_1 + 1)}.$$

(iii) For small delays $(T \simeq 0)$ the positive equilibrium e_+ again changes from stability to instability as the death rate η crosses the critical value $\eta_c = \dfrac{3(\mu\, m_1 + 1)R}{m_1(R - 1)}$.

For large delays $(T \simeq \infty)$ the positive equilibrium e_+ remains unstable.

Again we encounter a mathematical model (3.3) whose result is in contrast to the conjecture that only large delays cause instabilities. Here we see that instability is caused by small delays.

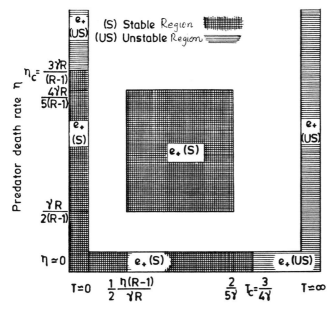

FIGURE C Crucial-egg-consumption-period T

Based on the results of Models A, B, and C , we summarize the possible responses for the stability of e_+ :

(a) e_+ may change its stability from stability to instability for increasing delays.

(b) e_+ may be stable or unstable for large delays.

(c) e_+ may be stable or unstable for small delays.

These responses do not allow us to give any definite answer to the basic question of this paper, "How would delays in predator's growth rate response affect the young predation stability results?" This non-committal situation further confirms the assertion made by Hastings (1983) that age-dependent predation is a complicated matter and that results about stability, limit cycles etc. are very model-dependent.

We conclude that increasing (decreasing) time delays are not necessarily destabilising (stabilising) confirming some earlier findings (Cushing (1982); Cushing and Saleem (1982)).

More importantly our results very clearly suggest that young predation with time delay is less stable than without it in the sense that with delays an otherwise stable positive equilibrium may be stable as well as unstable.

6 REFERENCES

Caperon, J., 1969. Time lag in population growth response of Isochrysis Galbana to a variable nitrate enviroment. Ecology, 50, 2: 188-192.

Caswell, H., 1972. A simulation study of a time lag population model. J. Theor. Biol., 34: 419-439.

Coleman, C.S. and Frauenthal, J.C., 1983. Satiable egg eating predators. Math. Biosci. 63: 99-119.

Cushing, J.M. 1977. Interodifferential equations and delay models in population dynamics. Lect. Notes in Biomath. 20, Springer-Verlag.

_____, 1982. Stability and maturation periods in age structured populations. Proc. Conf. on Diff. Eqn. in Biol. A.P.

_____ and Saleem, M. 1982. A predator prey model with age structure. J. Math. Biol. 14: 231-250.

Frauenthal, J.C. 1983. Some simple models of cannibalisms. Math. Biosci. 63: 87-98.

Freedman, H.I., and Rao, V.S.H., 1983. The trade-off between mutual interference and time lags in predator-prey systems. Bull. Math. Biol. 45: 991-1004.

Gopalsamy, K., 1984a. Delay responses and stability in two species systems. J. Austral. Math. Soc. Ser. B25: 473-500.

_____, 1984b. Harmless delays in a periodic ecosystem. J. Austral. Math. Soc. Ser. B25: 349-365.

Gurtin, M.E. and Levine, D.S. 1979. On predator-prey interactions with predation dependent on the age of prey. Math. Biosci. 47: 207-219.

Hastings, A., 1983. Age dependent predation is not a simple process I. Theory. Pop. Biol. 23: 347-362.

_____, 1984. Delays in recruitment at different tropic levels: Effects on stability. J. Math. Biol. 21: 1-35.

Levine, D.S., 1981. On the stability of a predator-prey system with egg-eating predators. Math. Biosci. 56: 27-46.

Lloyd, M. and Dybas, H.S., 1966. The periodical cicade problem. Evolution 20: 133-149; 466-505.

May, R.M., 1973. Time delays versus stability in population models with two and three trophic levels. Ecology 54: 315-325.

_____, 1974. Stability and Complexity in model ecosystems. 2nd ed. Princeton University Press, Princeton.

_____, Conway, G.R., Hassell, M.P., and Southwood, T.R.E., 1974. Time-delays, density dependence and single species oscillations, J. Anim. Ecol. 43, 3: 747-770.

McDonald, N., 1976. Time-delay in predator-prey models. Math. Biosci. 28: 321-330.

Nicholson, A.J., 1957. An outline of the dynamics of animal populations. Austral. J. Zool. 2: 9-65.

Nielsen, L., 1980. Effect of walleye (stizostedion Vitreum) predation on juvenile mortality and recruitment of yellow perch (perea flaveseens) in Oncida Lake, N.Y. Can. J. Fish. Aquat. Sci. 37: 11-19.

Saleem, M., 1983. Predator-prey relationships: egg eating predators. Math. Biosci. 65: 187-197.

_____, 1984. Egg eating age-structured predators in Interaction with age-structured prey. Math. Biosci. 70: 91-104.

_____, 1986. A mathematical model with young predation, preprint.

Shukla, V.P., 1983. Conditions for global stability of two species population models with discrete time-delay. Bull. Math. Biol. 45: 793-805.

Smith, J.M., 1974. Models in ecology. Cambridge University Press, Cambridge.

Smith, R.H. and Mead, R. 1974. Age structure and stability in models of predator-prey systems. Theor. Pop. Biol. 6: 308-322.

Solimano, F. and Beretta, E., 1983. Existence of a globally asymptotically stable equilibrium in Volterra models with continuous time-delay. J. Math. Biol. 18: 93-102.

Thompson, R.W., Dibiasio, D., and Charles, M., 1982. Predator-prey interactions: egg eating predators. Math. Biosci. 60: 109-120.

Volterra, V., 1927. Variazioni e fluttuazionidel numero d' individui in specie animali convivendi. R. Comit. Talars. Ital., Memoria 131, Venezia.

_____, 1931. Leions sur la theorie mathematique de la lutte pour la vio' Gauthiers - Villars, Paris.

UNIFORM PERSISTENCE AND GLOBAL STABILITY IN MODELS INVOLVING MUTUALISM 1. PREDATOR-PREY-MUTUALISTIC SYSTEMS

H.I. FREEDMAN*
Department of Mathematics, University of Alberta, Edmonton, Canada

BINDHYACHAL RAI+
Department of Mathematics, University of Allahabad, Allahabad, Uttar Pradesh, India

1 INTRODUCTION

In the literature dealing with the dynamics of the models in population ecology, most authors have concentrated on predator-prey or competitive interactions, while paying little or no attention to mutualistic interactions. Such early papers as existed dealt exclusively with two-species mutualism, while in recent years some work has been done on models involving a third species resulting in indirect mutualism.

The above mentioned papers deal with two types of mutualism, facultative, in which the mutualism is beneficial to both the species involved, but is not mandatory for the survival of the considered species, and obligate, which is mandatory for the species' survival.

It is our intention in this paper, after reviewing the work done on modeling both facultative and obligate mutualism in two-species models, to continue the analysis of three species models representing facultative mutualism with a prey (Rai, Freedman, and Addicott (1983)) and obligate mutualism with a predator (Freedman, Addicott, and Rai (in press)) . We will utilize the ideas of persistence (Freedman and Waltman (1984, 1985)) and uniform persistence (Butler, Freedman, and Waltman (1986)) to conclude when a positive equilibrium in the above-mentioned models exist. We then utilize a technique discussed in (Cheng, Hsu, and Liu (1981); Freedman and So (1985); Hsu (1978)) to construct a Liapunov function and derive criteria under which that equilibrium is globally asymptotically stable, and we illustrate this with an example.

We shall say that a population $N(t)$ such that $N(0) > 0$ persists if $\liminf_{t \to \infty} N(t) > 0$.

We say that $N(t)$ uniformly persists if $\liminf_{t \to \infty} N(t) \geq \delta > 0$ for all initial values $N(0) > 0$.

A system (uniformly) persists if all components (uniformly) persist.

The models in this paper will be given by systems of ordinary differential equations. We assume throughout sufficient smoothness so that solutions to initial value problems exist, are unique, and are continuable for all positive time.

*Research partially supported by The Natural Sciences and Research Council of Canada, Grant No. NSERC 4823.
+Research partially supported by V.G.C. Grant (India) No. F. 8-14/83 (SR-III)

2 TWO-SPECIES MODELS

The simplest two-species model of mutualism is given by the Lotka-Volterra system (Dean (1983); Freedman (1980); Goh (1980))

$$dx_1 dt = x_1(\alpha_1 - \beta_1 x_1 + \gamma_1 x_2)$$

$$, \qquad d/dt \qquad (2.1)$$

$$dx_2/dt = x_2(\alpha_2 - \beta_2 x_2 + \gamma_2 x_1)$$

$$x_i(0) = x_{i0} \geq 0, \ i = 1,2,$$

where $x_1(t)$ and $x_2(t)$ represent the mutualist population numbers at time t. The points $(0, 0)$, $(\alpha_1/\beta_1, 0)$, $(0, \alpha_2/\beta_2)$ are equilibria. Setting the right hand side equal to zero, we see that there is a positive equilibrium if and only if the isoclines given by the straight lines

$$\beta_1 x_1 - \gamma_1 x_2 = \alpha_1$$

$$(2.2)$$

$$\gamma_2 x_1 - \beta_2 x_2 = -\alpha_2$$

intersect in the interior of the first quadrant. If they intersect there, it is well known (Freedman (1980); Freedman, Addicott, and Rai (1982)) that this equilibrium so determined will be globally asymptotically stable with respect to all solutions such that $x_i(0) > 0$, $i = 1, 2$. If such a positive equilibrium does not exist, all solutions become unbounded, a biologically impossible situation, resulting in a bad model.

In (Albrecht, Gatzke, Haddad, and Wax (1974); Freedman (1980); Goh (1980); Kolmogorov (1936); Rescigno and Richardson (1967)) a more general model, the Kolmogorov model was considered. It is of the form

$$dx_1 dt = x_1 f_1(x_1, x_2), \ x_1(0) \geq 0$$

$$(2.3)$$

$$dx_2 dt = x_2 f_2(x_1, x_2), \ x_2(0) \geq 0,$$

where the assumptions

$$f_i(0, 0) > 0, \ i = 1, 2; \ f_1(K_1, 0) = f_2(0, K_2) = 0 \ \text{for some} \ K_1, K_2 > 0; \ \frac{\partial f_i}{\partial x_j}(x_1, x_2) > 0, \ j \neq i$$

hold. Under these assumptions it is shown in (Freedman, Addicott, and Rai (1983); Goh (1980)) that there is a globally asymptotically stable equilibrium.

All the above models are models of facultative mutualism. Models of two-species obligate mutualism have been considered in (Dean (1983); Freedman, Addicott, and Rai (1983); May (1976)). There it is shown that if the isoclines never intersect, all solutions approach zero, or if they intersect once, solutions either tend to zero or become unbounded, except for a one-dimensional manifold including the equilibrium. These are again bad models. However, if the isoclines intersect exactly twice in the positive octant, then one equilibrium is a saddle point, and the other is a node. Hence solutions generically either go to zero (extinction of the populations due to insufficient initial population numbers), or approach a stable-in-the-large equilibrium (the obligate mutualism taking effect).

In any case for all Kolmagorov models of two-species mutualism it is known (Freedman, Addicott, and Rai (1983); Goh (1980)) that provided $\partial f_i(x_1, x_2)/\partial x_i < 0$, $i = 1,2$, then there can be no nontrivial periodic solutions.

3 THE THREE-SPECIES MODEL

The system of equations

$$du/dt = uh(u, x, y) , u(0) \geq 0$$

$$dx/dt = xg(u, x) - yp(u, x) , x(0) \geq 0 \qquad\qquad (3.1)$$

$$dy/dt = y(-s(u, y) + c(u) p(u, x)) , y(0) \geq 0$$

was proposed as a model of facultative mutualism with a prey in (Rai, Freedman, and Addicott (1983) (there $s(u, y)$ and $c(u)$ were taken as constant) and of obligate mutualism with a predator in Freedman, Addicott, and Rai (1983) . Here $u(t)$, $x(t)$, $y(t)$ are the population numbers at time t of the mutualistic, prey, and predator populations respectively.

The hypotheses proposed in the above-mentioned papers designated respectively as Case I and II are:

Case I. (H1) $h(0, x, y) > 0$, $h_u(u, x, y) \leq 0$.

(H2) $h_x(u, x, y) > 0$, $h_y(u, x, y) \leq 0$.

(H3) $\exists L(x, y) > 0$, $h(L(x, y), x, y) = 0$.

(H4) $g(u, 0) > 0$, $g_x(u, x) \leq 0$.

(H5) $\exists K(u) > 0$, $g(u, K(u)) = 0$.

(H6) $p(u, 0) = 0$, $p_x(u, x) > 0$.

(H7) $p_u(u, x) \leq 0$.

(H8) $s_u(u, y) \geq 0$, $s_y(u, y) > 0$, $c'(u) \leq 0$.

(H9) $\lim_{x\to\infty} L(x, 0) = L^\circ < \infty$.

(H10) $s(0)/c(0) \in R(p, (0, x))$.

(H11) Let $p(0, x^\ddagger) = s(0)/c(0)$, then $x^\ddagger < K(0)$.

These hypotheses are slight modifications of those proposed in Rai, Freedman, and Addicott (1983), to take into account that here $s(u, y)$ and $c(u)$ are not necessarily constant. In addition we assume that one of the three inequalities (in which partial derivative with respect to u is involved) described in (H7, 8) is strict. The analysis in Rai, Freedman, and Addicott (1983) with only slight modifications carry over to this paper.

Case II: Hypothesis $H(1, 3, 4, 5, 6)$ hold. In addition we assume

(H2)' $h_x(u, x, y) \leq 0$, $h_y(u, x, y) > 0$.

(H7)' $p_u(u, x) \geq 0$.

(H8)' $s_u(u, y) \leq 0$, $s_y(u, y) \geq 0$, $c'(u) \geq 0$.

(H9)' $\lim_{y\to\infty} L(0, y) = L^\circ < \infty$.

(H10)' Either (H10) or (H11) is violated.

Interpretation of these various hypotheses are consistent with how the mutualistic, prey, and predator will grow separately and together in cases I and II respectively, and are described in details in Freedman, Addicott, and Rai (1983) and Rai, Freedman, and Addicott (1983) .

Models where the mutualist may be considered as a parameter and hence which can be reduced to two-species models are considered in Addicott and Freedman (1984) .

4 REGION OF ATTRACTION

It is relatively easy to show that all solutions of system (3.1) under the stated hypotheses are bounded for all positive time. The bounds are functions of the initial populations. However, for our results on global stability in section 7, we require an estimate of the region of attraction of the solutions of system (3.1).

In case I,

$u' = uh(u, x, y) \leq uh(u, x, 0) \Rightarrow 0 \leq \lim_{t\to\infty} u(t) \leq L^\circ$.

Then $x' = xg(u, x) - yp(u, x) \leq xg(u, x) \Rightarrow 0 \leq \lim_{t\to\infty} x(t) \leq \max_{0\leq u\leq L} K(u) \overset{\Delta}{=} K^\circ$. Now

consider $(c(u)x + y)' = c(u) xg(u, x) - s(u, y)y + c'(u)x \leq c(u) K^\circ g(u, 0) + s(u, y) c(u)x -$

$s(u, y)(c(u)x + y) \leq s(0, 0)(c(u)x + y) + c(0) K^\circ(g^\circ + s^\circ)$, where $g^\circ = \max\limits_{0 \leq u \leq L^\circ} g(u, 0)$ and

$\lim\limits_{y \to \infty} s(L^\circ, y) = s^\circ < \infty$. Then $0 \leq \lim\limits_{t \to \infty} (c(u)x + y)(t) \leq \dfrac{c(0)K^\circ}{s(0, 0)} (g^\circ + s^\circ)$, i.e.

$0 \leq \lim\limits_{t \to \infty} y(t) \leq \dfrac{c(0)K^\circ}{s(0, 0)} (g^\circ + s^\circ)$. Hence if we define

$A = \{(u, x, y): 0 \leq u \leq L^\circ , 0 \leq x < K^\circ , 0 \leq y \leq \dfrac{c(0)K^\circ}{s(0, 0)} (g^\circ + s^\circ)\}$, then the region of

attraction for solutions of system (3.1) is contained in A .

In case II, we again have $0 \leq u(t) \leq L^\circ$. If we define $K^\circ = \max\limits_{0 < u \leq K^\circ} K(u)$ by similar

arguments as in case I, we get that the region of attraction is contained in B , where

$B = \{(u, x, y): 0 \leq u \leq L^\circ , 0 \leq x \leq K^\circ , 0 \leq y \leq \dfrac{K^\circ(s^\circ c(L^\circ) + c(L^\circ)g^\circ + D)}{s(L, 0)}$

where $s^\circ = \lim\limits_{y \to \infty} s(0, y)$, $g^\circ = \max\limits_{0 \leq u \leq L^\circ} g(u, 0)$, $D = \max\limits_{0 \leq u \leq L^\circ} c'(u)$

5 PERSISTENCE AND EXTINCTION IN FACULTATIVE MUTUALISM WITH THE PREY

We now suppose the hypotheses of case I hold. We have analyzed a model very similar to this (c(u) and s(u, y) constant) in Rai, Freedman, and Addicott (1983), and shown that a variety of behaviors is possible. We have obtained general criteria for Hopf bifurcation and periodic solutions, and have shown in a specific case that the predator population can become extinct.

In this paper, we give general criteria for the uniform persistence of all three populations as well as general criteria for the extinction of the predator population.

As pointed out in Rai, Freedman, and Addicott (1983), equilibria on the coordinate planes are $E_0(0, 0, 0)$, $E_1(L(0, 0), 0, 0)$, $E_2(0, K(0), 0)$, $E_3(u^\circ, x^\circ, 0)$, $E_4(0, x^\ddagger, y^\ddagger)$. Let λ° be the eigenvalue of E_3 in the y-direction, and $\lambda^\ddagger$ be the eigenvalue of E_4 in the u-direction. Then by computing the variational matrix about E_3 and E_4 , we get

$$\lambda^\circ = -s(u^\circ, 0) + c(u^\circ) p(u, x^\circ)$$

$$\lambda^\ddagger = h(0, x^\ddagger, y^\ddagger) \tag{5.1}$$

Note that $\lambda^\ddagger > 0$ always exists.

The subsystem representing dynamics in the x - y plane has an equilibrium at E_4 . It may also have periodic orbits lying in that plane. Although we can handle that case, for simplicity we assume (H12) E_4 is asymptotically stable in the large with respect to solutions starting in the interior of the positive quadrant of the x - y plane.

Conditions for this to be valid are given in Cheng, Hsu, and Liu (1981) and Hsu (1978).

We can now state the persistence theorem for case I.

Theorem 5.1. Let (H1-12) hold. Then if $\lambda° > 0$, system 5.1 uniformly persists.

Proof. If $\lambda° > 0$, the technique used in Freedman and Waltman (1984, 1985) to obtain persistence precisely gives persistence in this case. Then by the results contained in Butler, Freedman, and Waltman (1986), persistence implies uniform persistence.

Note that by Butler, Freedman, and Waltman (1986), uniform persistence implies the existence of a positive equilibrium. Call it $E^*(u^*, x^*, y^*)$.

We now wish to consider criteria which lead to the extinction of the predator due to the action of the mutualist. To do this we must obtain a positive lower bound to the eventual mutualist population. We suppose

$$(H13) \quad \lim_{y\to\infty} L(x, y) = L^0(x) > 0 .$$

Let $L_0 = L^0(K°)$. Then $L_0 \leq \lim_{t\to\infty} u(t) \leq L°$.

Theorem 5.2. Let H(1-11, 13) hold. Then if

$$-s(L_0, 0) + c(L_0)\, p(L_0, K°) < 0 , \tag{5.2}$$

$$\lim_{t\to\infty} y(t) = 0 .$$

Proof. $y' = y(-s(u, y) + c(u)\, p(u, x)) \leq y(-s(L_0, 0) + c(L_0)\, p(L_0), K°))$ for sufficiently large t, say T. Hence by (5.2), the theorem follows.

6. PERSISTENCE IN OBLIGATE MUTUALISM WITH A PREDATOR

We now consider case II. Because of (H10)', E_4 no longer exists and in the absence of mutualism, the predator population goes to extinction. However a theorem quite similar to Theorem 5.1 can be shown in this case as well.

Theorem 6.1. Let H(1, 3-6, 2', 7'-10') hold. If $\lambda° > 0$, then system (3.1) uniformly persists, and has a positive equilibrium.

The proof is similar to the proof of Theorem 5.1, and the remark following it.

Theorem 6.1 can be interpreted biologically as follows. If a predator is unable to survive on a given prey on its own, then mutualism could help the predator population to survive by helping the prey population to grow to a higher level, by increasing the "hunting" efficiency of the predator, by supplying an additional food source, by increasing the efficiency of converting the predator's food to its own biomass, or by some combination of the above.

7 GLOBAL STABILITY OF THE EQUILIBRIUM

Under stated hypotheses in case I and II, we see that if $\lambda > 0$, system (3.1) persists and the positive equilibrium E^* exists. We now wish to obtain a criterion under which E^* is globally asymptotically stable with respect to $\text{Int}(R^3_+)$, the interior of the nonnegative octant.

Define $V(u, x, y)$ by

$$V(u, x, y) = u - u^* - u^* \ln(\frac{u}{u^*}) + \int_{x^*}^{x} \frac{-s(u^*, y^*) + c(u^*) p(u^*, \xi)}{p(u^*, \xi)} d\xi + y - y^* - y^* \ln(\frac{y}{y^*}). \quad (7.1)$$

$V(u, x, y)$ is a positive definite function about E^*. $V(u, x, y) \to \infty$ if either of $u, x,$ or $y \to \infty$ or for finite (u, x, y) if either of $u, x, y \to 0$.

We now compute $dV/dt(u, x, y)$, the derivative of V along solutions of system (3.1).

$$dV/dt(u, x, y) = (u - u^*) h(u, x, y) + [-s(u^*, y^*) + c(u^*) p(u^*, x)] \cdot [\frac{xg(u, x)}{p(u^*, x)} - \frac{yp(u, x)}{p(u^*, x)}]$$

$$+ (y - y^*)[-s(u, y) + c(u) p(u, x)] . \quad (7.2)$$

After some algebraic manipulations, (7.2) can be written as

$$dV/dt(u, x, y) = a_{11} + a_{12} + a_{13} + a_{22} + a_{23} + a_{33} , \quad (7.3)$$
where

$$a_{11} = (u - u^*) h(u, x^*, y^*)$$

$$a_{12} = (u - u^*) [h(u, x, y^*) - h(u, x, y)] + [-s(u^*, y^*) + c(u^*) p(u^*, x)]$$

$$[y - \frac{xg(u^*, x)}{p(u^*, x)} - \frac{yp(u, x)}{p(u^*, x)} + \frac{xg(u, x)}{p(u^*, x)}]$$

$$a_{13} = (u - u^*)]h(u, x, y) - h(u, x, y^*)] + (y - y^*)[s(u^*, y) - s(u, y) - c(u^*) p(u^*, x) + c(u) p(u, x)]$$

$$a_{22} = [-s(u^*, y^*) + c(u^*) \, p(u^*, x)] \, [\frac{xg(u^*, x)}{p(u^*, x)} - y^*] \tag{7.4}$$

$$a_{23} = 0$$

$$a_{33} = (y - y^*) \, [s(u^*, y^*) - s(u^*, y)] \, .$$

We can now write the a_{ij}'s as follows, defining functions b_{ij},

$$a_{11} = -b_{11}(u) \, (u - u^*)^2 \, ,$$

$$a_{12} = -2b_{12}(u, x, y)(u - u^*) \, (x - x^*) \, ,$$

$$a_{13} = -2b_{13}(u, x, y) \, (u - u^*)(y - y^*) \, ,$$

$$a_{22} = -b_{22}(x) \, (x - x^*)^2 \, ,$$

$$a_{33} = -b_{33}(y)(y - y^*) \, ,$$

$$b_{23} = 0 \, ,$$

$$b_{ji} = b_{ij} \, . \tag{7.5}$$

If we now define the 3×3 matrix $B(u, x, y)$ to be the matrix whose ij^{th} component is b_{ij}, then the following theorem is clear.

<u>Theorem 7.1.</u> Let $B(u, x, y)$ <u>be a positive definite matrix for all points in the set</u> $\{(u, x, y): (u, x, y) \in A\}$. <u>Then</u> E^* <u>is globally asymptotically stable with respect to all points in</u> Int R^3_+.

<u>Proof.</u> Let $(x_0, y_0, z_0) \in$ Int R^3. Let O be the orbit through this point, and Ω be the omega limit set of this orbit. Persistence implies that $\Omega \cap \{(u, x, y): u = 0$ or $x = 0$ or $y = 0\} = \emptyset$. Further the only invariant sets on ∂A are in $\{(u, x, y): u = 0$ or $x = 0$ or $y = 0\}$. Since A is an attracting set, $\Omega \subset A$ and by the above, $\Omega \subset$ Int A. But if B is positive definite for points in A, then dV/dt is negative definite, and by Liapunov's theorem $\Omega = E^*$, proving the theorem.

We note that the theorem can only be satisfied if $b_{11}(u) > 0$, $b_{22}(x) > 0$, $b_{33}(y) > 0$ for $(u, x, y) \in$ Int A. From our hypotheses $b_{11}(u) > 0$, $b_{33}(y) > 0$ are automatically satisfied. However, the condition $b_{22}(x) > 0$ puts some restrictions on the functions $g(u^*, x)$ and $p(u^*, x)$. Namely, we require

$$(x - x^*) \left[\frac{xg(u^*, x)}{p(u^*, x)} - y^*\right] < 0 \text{ for } 0 < x < K^\circ, x \neq x^*.$$

A geometrical interpretation of this condition can be given as follows. Consider the curve in the positive octant of the x - y plane given by $y = \frac{xg(u^*, x)}{p(u^*, x)}$. Generally this curve is continuous, passes through the points (in the x - y plane) $(0, \frac{g(u^*, 0)}{p_x(u^*, 0)})$, (x^*, y^*), $(K(u^*), 0)$, and may have several maxima and minima. Then $b_{22}(x) > 0$ can only be satisfied if x^* lies to the right of all local maxima (if any), and y^* lies below all local minima (if any).

Additional restrictions can now arise because of the mixed terms. We illustrate this with an example in the next section.

8 AN EXAMPLE

To illustrate the above analysis we consider the following example of case I, where the coefficients are taken for mathematical convenience, not necessarily from a real biological system.

$$du/dt = u(1 - \frac{u}{1 + x})$$

$$dx/dt = x(3 - x) - xy \tag{8.1}$$

$$dy/dt = y(-\frac{(1 + u)}{4} - y + x).$$

Equilibria are $E_0(0, 0, 0)$, $E_1(1, 0, 0)$, $E_2(0, 3, 0)$, $E_3(3, 2, 0)$, $E_4(0, \frac{13}{8}, \frac{11}{8})$, $E_5(3, 2, 1)$.

The eigenvalue of E_3 in the y direction is 1 and the eigenvalue of E_4 in the u direction is also 1. Hence the criteria for uniform persistence are satisfied.

Letting

$$V(u, x, y) = u - 3 - 3 \ln(\frac{u}{3}) + \int_2^x (-\frac{(1 + u)}{4} - y + \xi)\xi^{-1} d\xi + y - 1 - \ln y, \tag{8.2}$$

then

$$dV/dt(u, x, y) = -\frac{1}{3}(u - 3)^2 - \frac{1}{4}(u - 3)(y - 1) - (x - 2)^2 - (y - 1)^2 \tag{8.3}$$

which is negative definite for all (u, x, y) in the positive octant, and hence E_5 is globally asymptotically stable.

9 DISCUSSION

We have surveyed two-species mutualism models of both facultative and obligate types.

We have considered models of predator-prey-mutualist systems, in the case where the mutualism is with the prey and then the case of mutualism with the predator. We have given criteria in both cases for persistence of all three populations.

In such cases we have pointed out that an interior equilibrium exists, and have a given criteria for such an equilibrium to be globally asymptotically stable.

10. REFERENCES

Addicott, J.F. and Freedman, H.I., 1984. On the structure and stability of mutualist systems: Analysis of predator-prey and competition models as modified by the action of a slow growing mutualist, Theor. pop. Biol. 26: 320-339.

Albrecht, F., Gatzke, H., Haddad, A., and Wax, N., 1974. The dynamics of two interacting populations, J. Math. Anal. Appl., 46: 658-670.

Butler, G.J., Freedman, H.I. and Waltman, P., 1986. Uniformly persistent systems, Proc. Amer. Math. Soc., 96: 425-430.

Cheng, K.-S., Hsu, S.-B., and Liu, S.-S., 1981. Some results on global stability of a predator- prey system, J. Math. Biol. 12:115-126.

Dean, A.M., 1983. A simple model of mutualism, Am. Nat. 121: 409-417.

Freedman, H.I., 1980. Deterministic Mathematical Models in Population Ecology, Marcel Dekker, New York.

Freedman, H.I., Addicott, J.F., and Rai, B., 1983. Nonobligate and obligate models of mutualism, in Population Biology Proceedings, Edmonton 1982 (H.I. Freedman and C. Strobeck, eds)., Springer Verlag, Heidelberg, pp. 349-354.

_____, Obligate mutualism with a predator: Stability and persistence of three-species models, Theor. Pop. Biol. (in press).

Freedman, H.I. and So, J.W.-H., 1985. Global stability and persistence of simple food chains, Math Biosci. 76: 69-86.

Freedman, H.I. and Waltman, P., 1984. Persistence in models of three interacting predator-prey populations, Math. Biosci. 68: 213-231.

_____, 1985 . Persistence in a model of three competitive populations, Math. Biosci. 73: 89-101.

Goh, B.-S., 1980. Management and Analysis of Biological Populations, Elsevier Sci. Publ. Co., New York.

Hsu, S.-B., 1978. On global stability of a predator-prey system, Math. Biosci. 39: 1-10.

Kolmogorov, A.N., 1936. Sulla teoria di Volterra totta per l'esisttenza, Gior. Instituto Ital. Attuari 7: 74-80.

May, R.M., 1976. Models for two interacting populations, in Theoretical Ecology (R.M. May, ed.) Saunders Publ. Co., Philadelphia, pp. 49-70.

Rai, B., Freedman, H.I. and Addicott, J.F., 1983. Analysis of three species models of mutualism in predator-prey and competitive systems, Math Biosci. 63: 13-50.

Rescigno, A., and Richardson, I.W., 1967. The struggle for life. I: Two species, Bull. Math. Biophys. 29: 377-388.

PART V - RESOURCE MANAGEMENT

DYNAMIC INTERACTIONS BETWEEN ECONOMIC, ECOLOGICAL AND
DEMOGRAPHIC VARIABLES

ALESSANDRO CIGNO, Department of Economics and Commerce, University of Hull,
Hull HU6 7RX ENGLAND

1 INTRODUCTION

The present paper is intended as an introduction to models of the interaction
between economic, ecological, and demographic variables. The history of this literature
reflects, to some extent, the changing public perceptions of the importance of natural
constraints on economic development. Thus, the optimism of the post-war years
produced models like the one illustrated in the first section of this paper, where the rise
in living standards is restricted only by man's technological inventiveness. Evidence of
natural resource scarcity from the beginning of the seventies (rising prices of oil and
other primary commodities, growing awareness of environmental decay in heavily
industrialized areas, slower economic growth) prompted a series of investigations into
the possible consequences of finite natural resource constraints and gave rise to some
alarmist predictions. By contrast, the more mature models of the sober eighties show
that living standards need not fall and may be enhanced under certain conditions.

For reasons of brevity and clarity, the exposition is based on a highly stylized and
easily manipulable representation of reality. The results presented are, therefore,
suggestive rather than conclusive, but the main propositions stand up to more general
treatment. Throughout the exposition, I shall emphasize the existence and stability of
indefinitely sustainable development paths.

2 A SIMPLE ECONOMIC MODEL

A suitable starting point for this exposition is the Solow-Swan model of an
economy producing a good Y by means of the services of Capital K and
Labour L, where capital is defined as the accumulated stock of goods produced in the
past but not consumed. In reality economies produce many different commodities,
some of which can be consumed (like food), while others (like industrial machinery) can
only be used for further production, but it simplifies matters considerably to think of Y
as a homogeneous and perfectly malleable substance, which can be instantly and
costlessly molded into any desired combination of consumer's and producer's goods.

2.1. Given Technology

The centrepiece of this model is the equation

$$Y = F(K, L) , \qquad\qquad (1)$$

where $F(K, L)$ is the <u>production function</u>, describing all efficient combinations of inputs and output permitted by existing technology. This function is customarily assumed to have a number of properties. First, both factors are essential to production,

$$F(0, L) = F(K, 0)) \equiv 0 . \tag{2}$$

Second, the technology displays constant returns to scale (linear homogeneity)

$$F(\lambda K, \lambda L) = \lambda F(K, L) \text{ for any } \lambda > 0 , \tag{3}$$

implying that "nature" does not impose a constraint on production. Third, it is at least twice differentiable, with

$$F_i > 0 \text{ and } F_{ii} < 0 \ (i = K, L) \text{ for } (K, L) > 0 , \tag{4}$$

meaning that each factor's marginal contribution to production, holding the other factor constant, is always positive (by the efficiency assumption) but decreasing (i.e. the factor held constant does impose a constraint on production). Given the expository purpose of this paper, it will be convenient to work throughout with a particular form of the production function,

$$F(K, L) \equiv K^{\alpha_1} L^{\alpha_2} , \alpha_i > 0 , \Sigma_i \alpha_i = 1 , \tag{5}$$

which satisfies all the above assumptions, even though much of what follows has been proved using unspecified functions and is thus of more general validity.

The model is completed by stipulating

$$\dot{K} = sY , 0 < s < 1 , \tag{6}$$

and

$$\dot{L} = nL \tag{7}$$

where s and n are behavioural-institutional parameters representing, respectively, the fraction of output "saved" (i.e. added to the stock of capital) and the rate of population growth (assuming a constant rate of labour force participation).

Given (3), it is clear that we can describe the state of the system at any instant by the ratio of any two of the variables K, L, and Y. In view of subsequent extensions to the model, I shall opt for the output/capital ratio $\beta \equiv Y/K$. The fundamental dynamic equation of the system is then

$$\dot{\beta} = \alpha_2(n - s\beta)\beta , \tag{8}$$

which has a unique stationary solution

$$\beta^* \equiv n/s . \tag{9}$$

At β^* the system is in a steady state, because K, L, and Y are all growing at a constant proportional rate n .

Since β must be positive, otherwise the economy would vanish, it is clear that

$$n > 0 \tag{10}$$

is necessary and sufficient for the existence of a steady state. If such a state exists, it will also be globally asymptotically stable, because

$$\dot{\beta} \gtrless 0 \text{ as } \beta \lessgtr \beta^* \tag{11}$$

Provided that the population is expanding, the system will thus converge from any arbitrary initial position to a steady state characterized by an output/capital ratio constant in time. The steady-state output/labour ratio, $y \equiv Y/L$, is also constant in time, but decreasing in n ,

$$y^* = (s/n)^{\alpha_1 \alpha_2} . \tag{12}$$

Since consumption is proportional to output, and output per head is proportional to output per unit of labour, (12) faces society with a harsh choice between having more children and raising living standards: the higher the rate of population growth, the lower the asymptotic level of per-capita consumption, forever.

2.2 Technological progress

The analysis of the previous section can be extended to allow for improvements in technology. These are customarily modelled by making output an increasing function of calendar time, t , as well as of K and L ,

$$Y = K^{\alpha_1} , L^{\alpha_2} e^{mt} , \alpha_i > 0 , \Sigma_i\alpha_i = 1 , m > 0 \tag{13}$$

The differential equation for β then becomes

$$\dot{\beta} = a_2[\frac{m + \alpha_2 n}{\alpha_2} - s\beta]\beta \, , \tag{14}$$

which has a unique stationary solution,

$$\beta^* \equiv \frac{m + \alpha_2 n}{\alpha_2 s} \, , \tag{15}$$

if and only if

$$n > -m/\alpha_2 \, . \tag{16}$$

If such a solution exists, it is globally stable as in the basic model.

An important effect of technical progress is thus to permit a stable steady state even in the event that the population is stationary or declining (but not too fast). Another is that the asymptotic output-labour ratio is no longer constant in time,

$$y^* = [\frac{\alpha_2 \, s}{m + \alpha_2 n}]^{(\alpha_1/\alpha_2)} e^{(m/\alpha_2)t} \tag{17}$$

because in the steady state, total output (and capital) grows at the rate $(m + n)$, while labour grows at the rate n. Notice, however, that a rise in n would still lower the time-profile of y^* as in the case of fixed technology.

An unrealistic feature of (13) is that the rate of technical progress is exogeneously determined and thus independent of past history. A less crude approach, originally proposed in Arrow (1962) and recently developed in Cigno (1984a), is to postulate that personal skills and technical knowledge are the result of "experience", as measured by the total amount of output ever produced, Q. Staying with the same functional form (and preserving linear-homogeneity in (K, L)), we can then write

$$Y \equiv \dot{Q} = K^{\alpha_1} L^{\alpha_2} Q^\mu \quad , \alpha_i > 0 \, , \Sigma_i \alpha_i = 1, \ 0 < \mu < 1 \, . \tag{18}$$

This makes the rate of technical progress endogeneous and proportional to the growth rate of cumulative output, $\delta \equiv Y/Q$.

The economy is now governed by the dynamic system,

$$\dot{\beta} = -\alpha_2 s\beta^2 + \mu\beta\delta + \alpha_2\beta n \tag{19}$$

$$\dot{\delta} = \alpha_1 s\beta\delta - (1 - \mu)\delta^2 + \alpha_2 \, \delta n \, , \tag{20}$$

which has a unique stationary solution,

$$\beta^* \equiv \frac{\alpha_2}{\alpha_2 - \mu} \frac{n}{s} , \delta^* \equiv s\beta^* , \qquad (21)$$

if and only if

$$(\alpha_2 - \mu)n > 0 . \qquad (22)$$

Such a solution will be globally asymptotically stable (see Figure 1) if and only if

$$\mu < \alpha_2 , \qquad (23)$$

which in view of (22) implies $n > 0$.

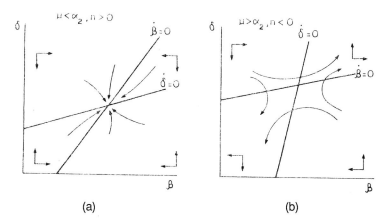

(a)   (b)

FIGURE 1 Phase diagrams for (β, δ)-system

       An expanding population is thus required for stability in the face of endogeneous technological change. If the population were contracting, a steady state could still exist, but any external shock could either send the economy to extinction in finite time, or set off a process of accelerating growth (see Figure 1b). The latter might be thought to be a desirable scenario but, apart from the fact that it could be reversed by any subsequent shock, it stretches beyond credibility the assumption that "nature" does not impose a constraint on the productive processes.

       The steady-state output/labour ratio,

$$y^* = Y_0{}^{(\mu/\alpha_2)} s^{(\alpha_1/\alpha_2)} [\frac{\alpha_2 - \mu}{\alpha_2 n}]^{\{(\alpha_1 + \mu)/\alpha_2\}} e^{\{\mu n/(\alpha_2 - \mu)\}} \qquad (24)$$

is still increasing in time as in the exogeneous technical progress formulation, but, assuming that the stability condition (23) is satisfied, a rise in n will now rotate the time-profile of y* anticlockwise (as in Figure 3a). Thus the consumption-procreation trade-off noted in the exogeneous technical progress case applies here only to the earlier generations.

3 NATURAL RESOURCES AND POPULATION CONTROL

Let us now turn to the case where economic activity depletes some natural resource.

3.1. Exogenous technical progress

A model presented in Stiglitz (1974) assumes

$$Y = K^{\alpha_1} L^{\alpha_2} R^{\alpha_3} e^{mt} , \quad \alpha_i > 0 , \ \Sigma_i \alpha_i = 1 , \ m > 0 , \tag{25}$$

subject to

$$\int_0^\infty R \, dt \leq S_0 . \tag{26}$$

where S_0 is the initial stock of a natural resource which is necessarily used up or polluted in the productive process at a (variable) rate R .

Notice that the technology is assumed to allow some degree of substitutability of K and L for R , either in the literal sense that any desired amount of Y can be produced by an input mix which contains less of the natural resource and more of capital and labour, or in the sense that some capital and labour can be diverted from production proper to pollution control. Notice, also, that the production function is now homogeneous of degree one in (K, L, R) , but of degree less than one (decreasing returns to scale) in (K, L) .

Let us now suppose that the stock of the natural resource is owned by a large number of income maximising individuals and that production takes place in a large number of profit maximising firms. Economic theory tells us that, in equilibrium, the following will be true at each instant:

(i) the natural resource is used up to the point where its marginal contribution to production (the first partial derivative of the production function with respect to R) is equal to the price of the resource in terms of output, p ,

$$\alpha_3 Y/R = p , \tag{27}$$

(ii) The services of capital are used up to the point where their marginal contribution to production is equal to their price (the rental price of capital) in terms of output, r,

$$\alpha_1 Y/K = r, \tag{28}$$

(iii) The rates of return to holding a stock of the natural resource (the rate of appreciation in its market value) and to holding capital (its rental price) are the same,

$$\dot{p}/p = r. \tag{29}$$

The same would be true if the resource were collectively owned <u>and</u> its use efficiently regulated.

Ignoring for the moment the feasibility constraint (26), we can boil (6)-(7)-(25)-(27)-(28)-(29) down to

$$\dot{\beta} = -\frac{\alpha_1\alpha_3 + \alpha_2 s}{\alpha_1 + \alpha_2}\beta^2 + \frac{m + \alpha_2 n}{\alpha_1 + \alpha_2}\beta \tag{30}$$

and

$$\dot{\gamma} = \gamma^2 + \frac{m - \alpha_1(1-s)\beta + \alpha_2 n}{\alpha_1 + \alpha_2}\gamma, \tag{31}$$

where $\gamma \equiv R/S \equiv -\dot{S}/S$ is the proportional rate of resource depletion.

Since γ, like β, must obviously be positive, (30)-(31) has a unique stationary solution

$$\beta^* \equiv \frac{m + \alpha_2 n}{\alpha_1\alpha_3 + s\alpha_2}, \gamma^* \equiv (\alpha_1 - s)\beta^* \tag{32}$$

if and only if

$$n > -m/\alpha_2 \text{ and } s < \alpha_1. \tag{33}$$

The presence of an exhaustible or pollutable resource has thus the effect of setting an upper limit $(\alpha_1 Y)$ to the rate of saving that the economy can absorb. More crucially, the combination of natural resource depletion with <u>exogenous</u> technical progress makes the steady state globally and locally unstable for all possible values of

212

the parameters (see Cigno (1981)). As shown in Figure 2, (β^*, γ^*) is a saddle-point; all trajectories other than the two which lead to it will either hit the β axis or violate the feasibility constraint (26) in finite time. Since both capital and the natural resource are essential to production, the chances of survival for such an economy are not very good!

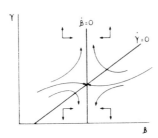

FIGURE 2 Saddle point configuration

3.2. Endogenous technical progress

The endogenous technical progress version of the exhaustible natural resource model is in Cigno (1984a). The production equation for this case may be written as

$$Y \equiv \dot{Q} = K^{\alpha_1} L^{\alpha_2} R^{\alpha_3} Q^{\mu} , \ \alpha_i > 0 , \ \Sigma_i \alpha_i = 1 , \ 0 < \mu < 1 , \tag{34}$$

which combines with (6), (7), (27), (28) and (29) to yield

$$\dot{\beta} = - \frac{\alpha_1 \alpha_3 + \alpha_2 s}{\alpha_1 + \alpha_2} \beta^2 + \frac{\mu}{\alpha_1 + \alpha_2} \beta\delta + \frac{\alpha_2}{\alpha_1 + \alpha_2} \beta n , \tag{35}$$

$$\dot{\gamma} = - \frac{\alpha_1 (1 - s)}{\alpha_1 + \alpha_2} \beta\gamma + \gamma^2 + \frac{\mu}{\alpha_1 + \alpha_2} \gamma\delta + \frac{\alpha_2}{\alpha_1 + \alpha_2} \gamma n \tag{36}$$

and

$$\dot{\delta} = \frac{\alpha_1 (s - \alpha_3)}{\alpha_1 + \alpha_2} \beta\delta + \frac{\mu - \alpha_1 - \alpha_2}{\alpha_1 + \alpha_2} \delta^2 + \frac{\alpha_2}{\alpha_1 + \alpha_2} \delta n . \tag{37}$$

This system has a unique stationary solution,

$$\beta^* \equiv \frac{\alpha_2\, n}{\alpha_1\alpha_3 + s(\alpha_2 - \mu)} \,,\, \gamma^* \equiv (\alpha_1 - s)\beta^* \,,\, \delta^* \equiv s\beta^* \tag{38}$$

if and only if

$$s < \alpha_1 \text{ and } \frac{\alpha_1 - s}{s} \gtrless \frac{\mu - \alpha_2 - \alpha_3}{\alpha_3} \text{ as } n \gtrless 0 \,,\, n \neq 0. \tag{39}$$

(notice that (38)-(39) is equivalent to (32)-(33) for $\mu = m = 0$). As in the exogenous technical progress case, the existence of a steady state is thus conditional upon the fraction of output saved and re-invested not exceeding a technologically determined ceiling. But, if the population is expanding, there is now also a floor below which saving must not fall. As in the endogenous technical progress model without natural resources, the population level is not allowed to stand still.

The advantage of endogenous over exogenous technical progress in the presence of natural resources is that there are conditions (see Cigno (1984a) for the derivation) under which the steady state is at least locally stable:

$$\frac{\alpha_1 - s}{s} < \frac{1 + \alpha_2 - \mu}{1 - 2\alpha_3}\,, \tag{40}$$

$$\frac{\alpha_1 - s}{s} < \frac{\alpha_1\alpha_3 + (\alpha_2 - \mu)s}{\alpha_1\alpha_3 + (\alpha_1 + 2\alpha_2 - \mu)s} \tag{41}$$

and

$$\frac{\alpha_1 - s}{s} < \frac{\mu - \alpha_2 - \alpha_3}{\alpha_3}\,. \tag{42}$$

The third of these conditions implies, for (39), that the rate of population growth must be negative. The second is automatically satisfied if (42) is. Provided that the population is declining, a steady state will thus exist and be stable if $\{(\alpha_1 - s)/s\}$ is positive and does not exceed the smaller of the two expressions, $\{(1 + \alpha_2 - \mu)/(1 - 2\alpha_3)\}$ and $\{(\mu - \alpha_2 - \alpha_3)/\alpha_3\}$. Since s can be modified, directly or indirectly, by government action, this means that the system can actually be steered on a sustainable path and that, once there, it would be able to withstand small disturbances.

Although valid only for declining populations, this optimistic conclusion encourages one to look further into the properties of the steady state. The associated time-path of per-capita income and output is given by

$$y^* \equiv \{(\alpha_1 - s)(S_0/L_0)\}^{(\alpha_3)/(\alpha_2 + \alpha_3)}\{\frac{\alpha_2\, n}{\alpha_1\alpha_3 + s(\alpha_2 - \mu)}\}^{(\alpha_3 - \alpha_1)/(\alpha_2 + \alpha_3)}$$

$$Q_0^{\mu/(\alpha_2+\alpha_3)} e^{-\{(\alpha_1\alpha_3-\mu s)/[\alpha_1\alpha_3+(\alpha_2-\mu)s]\}(nt)} \tag{43}$$

Here the first thing to be noted is that y^* is increasing in (S_0/L_0): the greater the initial resource endownment per person the higher the material standard of living. Disregarding the unstable case, we also find that y^* is increasing in t, and that a rise in n would rotate the time-profile of y^* clockwise if α_3 is less than α_1 (Figure 3b), or lower it if α_3 is more than α_1 (Figure 3c).

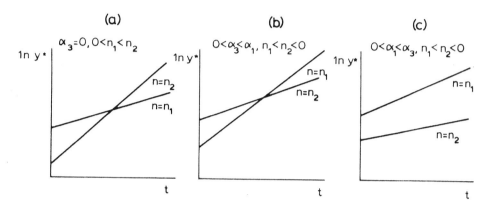

FIGURE 3 See text for description

The introduction of a natural resource constraint in the endogenous technical progress model has thus the effect of redistributing the conse- quences of higher n in favour of the earlier generations. If natural resources are sufficiently important $(\alpha_3 > \alpha_1)$, those consequences are negative for everyone (but worse for later generations).

4 ENDOGENOUS POPULATION CHANGE

The assumption made so far, that the rate of population change is unaffected by systematic changes in the standard of living and attendant changes in the structure of society, conflicts with the historical evidence. In this section we examine the implications of

$$n = y^{v_1} k^{v_2}, k \equiv K/L, \tag{44}$$

and derive the restrictions on v_i that would ensure a stable, sustainable equilibrium.

Although the choice of this particular functional form (similar to that of the production function) is purely a matter of expositional convenience, the existence of a relationship between the rate of population change, the level of output (or real income) per head and the degree of capital intensity of the production process is broadly consistent with economic theory, according to which n should be increasing in $y(v_1 > 0)$ and decreasing in $k(v_2 < 0)$ (see Cigno (1984b) for a fuller discussion).

4.1 Exogenous technical progress

This situation is examined in Cigno (1981). Suppose that the production function is as in (25). In addition to the differential equations (30) and (31), which remain valid, we now have

$$\dot{n} = -\frac{\alpha_1 v_1 + (\alpha_1 + \alpha_2)v_2}{\alpha_1 + \alpha_2} n^2 + \frac{mv_1 - \{\alpha_1 (\alpha_3 - s)v_1 - s(\alpha_1 + \alpha_2)v_2\}\beta}{\alpha_1 + \alpha_2} n . \tag{45}$$

A unique stationary solution,

$$\beta^* \equiv \frac{m}{\alpha_1 \alpha_3} , \gamma^* \equiv (\alpha_1 - s)\beta^* , n^* \equiv s\beta^* , \tag{46}$$

will exist so long as

$$m > 0 \text{ and } s < \alpha_1 \tag{47}$$

Therefore, the restriction on the rate of technical progress is stronger than in the exogenous population case. On the other hand, however, the steady state is now locally stable for a range of values of the parameters satisfying:

$$v_1 + v_2 < 0 , \tag{48}$$

$$\frac{\alpha_1 - s}{s} < \frac{1 - \alpha_1 (1 - v_1 - v_2) + \alpha_2 v_2}{1 - 2\alpha_3} , \tag{49}$$

$$\frac{\alpha_1 - s}{s} > -\frac{\alpha_1 \alpha_3 (v_1 + v_2)}{\alpha_1 (\alpha_3 + sv_1) - s(\alpha_2 + v_2)} . \tag{50}$$

Condition (48) is not too difficult to meet, because v_2 is expected to be negative (industrialisation raises, directly and indirectly, the opportunity cost of having children), and s can be chosen to satisfy (49)-(50).

Interestingly, the steady state behaviour of this model is rather like that of the basic model with exogenous population and no technical change or natural resources which we considered at the very start: the rate of population growth is positive and constant over time, as is the level of per-capita output:

$$y^* \equiv (\alpha_1 - s)^{\alpha_3/(\alpha_2 + \alpha_3)} (m/\alpha_1\alpha_3)^{(\alpha_3 - \alpha_1)/(\alpha_2 + \alpha_3)}$$

$$(S_0/L_0)^{\alpha_3/(\alpha_2 + \alpha_3)} \,, \tag{51}$$

because the rate of natural resource extraction or pollution is fine-tuned to be exactly compensated by technical progress. The only important difference is that y^* is now an increasing function of the initial per-capita endowment of the natural resource.

4.2. Endogenous technical progress

If the production function is as in (34), the relevant differential equations are (35), (36), (37) and (see Cigno (1984b) for the derivation)

$$\dot{n} = \frac{\alpha_1(s - \alpha_3)v_1 + s(\alpha_1 + \alpha_2)v_2}{\alpha_1 + \alpha_2}\, \beta n$$

$$+ \frac{\mu v_1}{\alpha_1 + \alpha_2}\, \delta n - \frac{\alpha_1 v_1 + (\alpha_1 + \alpha_2)v_2}{\alpha_1 + \alpha_2}\, n^2 \,. \tag{52}$$

For any arbitrary $\beta > 0$, this homogeneous system has a stationary solution of the form

$$n^* \equiv \delta^* \equiv s\beta \,, \quad \gamma^* \equiv (\alpha_1 - s)\beta \tag{53}$$

if and only if

$$0 < \frac{\alpha_1 - s}{s} = \frac{\mu - \alpha_3}{\alpha_3} \,. \tag{54}$$

As in the exogenous technical progress version, the steady-state level of per-capita output is constant in time and increasing in (S_0/L_0),

$$y^* \equiv \{(\alpha_1 - s)(S_0/L_0)\}^{\alpha_3/(\alpha_2 + \alpha_3)}\, \beta^{(\alpha_3 - \alpha_1)/(\alpha_2 + \alpha_3)}\, Q_0^{\mu/(\alpha_2 + \alpha_3)} \tag{55}$$

Again as in the previous version, steady states are locally stable for a range of values of the parameters:

$$\frac{\alpha_1 - s}{s} < \frac{\alpha_1 (2 + v_1 + v_2) + \alpha (1 + v_2)}{\alpha_1 + \alpha_2}, \tag{56}$$

$$\frac{\alpha_1 - s}{s} < 1 + \frac{\alpha_1 \mu (1 - v_1) - (\alpha_1 + \alpha_2)^2}{\alpha_1 (1 + v_1 + v_2) + \alpha_2 (2 + v_2)}, \tag{57}$$

$$\frac{\alpha_1 (v_1 + v_2) + \alpha_2 (1 + v_2)}{v_1 + v_2} < 0. \tag{58}$$

These conditions are in some respects less stringent than (48)-(49)-(50), because they do not prescribe the sign of $(v_1 + v_2)$, but, on the other hand, there is no guarantee that (58) can be met by a judicious choice of s. The existence conditions (54) are also rather stronger than (47).

5 CONCLUDING REMARKS

A general conclusion that can be drawn from the foregoing exercises is that economic, demographic and ecological variables interact over time in an essential way. In particular, the naive prediction of the basic economic model, that the economy will automatically seek a unique equilibrium path from whatever initial position and then stick to it for ever and ever is, at best, an over-simplification: when demographic and ecological feed-backs are taken into account, we find that the sustainable trajectory (or trajectories) depends on initial conditions and, furthermore, that the system may need to be steered towards it by deliberate policy. On the other hand, the more catastrophic predictions of the 1970's appear to be the result of omitted variables or inadequate representation of technical progress, and thus as unwarranted as the excessively optimistic ones of earlier years.

6 REFERENCES

Arrow, K.J., 1962. "The Economic Implications of Learning by Doing" in Review of Economic Studies, Vol. XXIX.
Cigno, A., 1981. "Growth with Exhaustible Resources and Endogenous Population" in Review of Economic Studies, Vol. XLVIII.
_____, 1984a. "Further Implications of Learning by Doing" in Bulletin of Economic Research, Vol. 36.
_____, 1984b. "Consumption vs. Procreation in Economic Growth" in Economic Consequences of Population Change in Industrialised Countries (G. Steinmann, ed.), Berlin, Springer-Verlag.
Solow, R.M., 1956. "A Contribution to the Theory of Economic Growth", in Quarterly Journal of Economics.
Stiglitz, J., 1974. "Growth with Exhaustible Natural Resources: the Competitive Economy" in Review of Economic Studies, Vol. XLIV.

ECONOMIC GROWTH MODELS: EFFECTS OF LOGISTIC POPULATION AND TECHNOLOGY

AJAI SHUKLA
National Technical Manpower Information System (Nodal Centre)
Indian Institute of Technology, Kanpur 208 016 INDIA

1 ABSTRACT

Economic growth models with logistic population and technology are studied in the cases of a plentiful resource and an exhaustible resource. In each instance the conditions under which the economy may follow a sustainable path have been derived.

2 INTRODUCTION

The dynamics of growth behaviour of an economic system depends upon numerous factors such as output, capital stock, labour force, population size, level of technology, quantum of resource, etc. and is mainly governed by the linear and nonlinear interactions of these factors in the model. During the last few decades extensive investigations have been carried out to analyse the dynamics of growth models of economic systems with plentiful as well as exhaustible resource under an exogeneous labour force, Solow (1956, 1974), Swan (1956), Uzawa (1961), Arrow (1962), Phelps (1966), Stiglitz (1974a,b), Suzuki (1976), Laitner (1984). Effects of an endogeneous labour population with plentiful and exhaustible resource have also been studied; Kelley (1974), Kosobud et. al. (1974), Schuler (1979), Stone (1980), Cigno (1981, 1984 a, b) . In particular Solow (1956) has shown that the capital-labour ratio remains finite when the exogeneous labour force grows exponentially, while rest of the economic variables grow without limit. Stone (1980) has studied Solow's (1956) model by considering a modified savings function and a labour population relation where the growth of the population is taken to be Gompertzian (1825) and the level of technology constant. He has shown that for a particular set of values of parameters not only the capital-labour ratio remains finite but also the other economic variables. Stiglitz (1974a, b) has shown that an exhaustible resource has the tendency to destablize an otherwise stable economic system which has an exponentially growing labour force. Cigno (1981), however, has proved that if the growth rate of labour force is taken to be a function of consumption and capital per capita, the unstable economic system with exhaustible resource may become stable for a wide range of values of savings - income ratios. The models of Stiglitz (1974 a, b) and Cigno (1981) have not taken into account the costs involved in the extraction of the resource and this aspect has been studied by Suzuki (1976) and Laitner (1984) by modifying the production and savings functions respectively.

In economic models with an exhaustible resource, the labour population relation, as considered by Stone (1980) should also endogeneously involve the initial resource endowment and resource used. These aspects may be studied by considering the logistic growth of population as assumed in demographic and ecological sciences (May (1973); Pielou (1977)). Similarly, the effect of change in the level of technology on production may also be studied by assuming that it follows logistic behaviour due to various economic contraints on capital, technical manpower, etc.

  We study the dynamic behaviour of economic models with a plentiful (no scarcity) or an exhaustible resource under the condition of a logistically growth population and technology. The labour force, assumed to be endogeneous, depends upon population, capital, output, total initial resource endowment and resource used. Under these conditions, criteria for both linear and non-linear stability of economic models are derived by using Liapunov's second method (Lasalle and Lefschitz (1961)).

3 MODEL WITH LOGISTIC POPULATION AND TECHNOLOGY: NO SCARCE RESOURCE

  Consider a single sector economy in which production of a single commodity takes place with the use of labour, capital and technology. Resource is abundant, in relation to demand, and hence any change in its stock would not affect economic growth. The behaviour of such an economy may then be assumed to be governed by the following factors (Stiglitz (1974 a, b); Cigno (1981, 1984 a, b; this volume)).

  (i) <u>Production Function</u> The production process in the economy is prescribed by the Cobb-Douglas function (Solow (1956, 1974), Schular (1979)):

$$Y = K^{\alpha_1} L^{\alpha_2} T^{\alpha} \tag{1}$$

$$\alpha_i \geq 0 \quad \Sigma \alpha_i = 1 \quad 0 \leq \alpha \leq 1$$

where,

    $Y(t)$ is the aggregate output, at time t ,
    $K(t)$ is the capital stock, at time t ,
    $L(t)$ is the labour force at time t , and
    $T(t)$ is the level of technology in use, at time t .

  (ii) <u>Capital Accumulation</u> The rate at which capital is growing is given by Stone (1980):

$$\frac{dK}{dt} = sY - s_0 K ; \quad s > 0 , s_0 \geq 0 \tag{2}$$

where s is a constant coefficient representing the fraction of the output used for capital growth and s_0 the constant fraction of capital lost due to depreciation.

(iii) Labour-Population Relation The supply of labour is assumed to be a variable function of the total population and depends on per capita output and capital per capita (Kosobud et al. (1974); Stone (1980); Cigno (1981, 1984, this volume)).

$$L = fP \tag{3}$$

where

$$f = \left(\frac{P}{K}\right)^{\mu_1} \left(\frac{Y}{P}\right)^{\mu_2}, \qquad \mu_i \geq 0 \tag{4}$$

μ_1, μ_2 are constants. The equation (3) reduces to Stones (1980) model when $\mu_2 = 0$.

(iv) Population Growth The rate of growth of population is taken to follow the Pearl- Verhurst logistic model:

$$P = n_0 (1 - \frac{P}{P_0}) P \tag{5}$$

where P is the population size, n_0 the rate of growth of population, P_0 the maximum sustainable population size under given environmental and ecological constraints. In this model population grows rapidly initially and then tends to taper off asymptotically to P_0 (Verhulst (1938); Pearl (1930); May (1973); Solow (1974); and Pielou (1977)).
When $P_0 \to \infty$, (4) gives the usual exponential growth of population. For $\alpha = 1$, $P_0 \to \infty$, $\mu_1 = 0$, $\mu_2 = 0$, (3), (4) reduce to the familiar exponentially growing labour force.

(v) Technical Change The level of technology used as an input in the production process may not always grow exponentially and may be constrained due to various impeding factors such as the non-availability of technical personnel, lack of research and development facility, low level of technical education. The equation governing the change in the technology T(t) in such a case can be written as

$$\frac{d}{dt} T = \lambda (1 - \frac{T}{T_0}) T , \tag{6}$$

where λ is the rate of growth and T_0 the maximum technology that could be available under various economic and other constraints.

3.1 Dynamic Equations

In this case we study the dynamics of β, $\beta = Y/K$, the output-capital ratio. From (1), (3) and (4) we get:

$$\frac{\alpha_1}{Y} \frac{dY}{dt} = \frac{\alpha_1}{K} \frac{dK}{dt} + \frac{\alpha_2}{L} \frac{dL}{dt} + \frac{\alpha}{T} \frac{dT}{dt} \tag{7}$$

and,

$$\frac{1}{L} \frac{dL}{dt} = (1 + \mu_1 - \mu_2) \frac{1}{P} \frac{dP}{dt} - \mu_1 \frac{1}{K} \frac{dK}{dt} + \mu_2 \frac{1}{Y} \frac{dY}{dt} \tag{8}$$

On substituting (8) in (7), we get,

$$\frac{1}{\beta} \frac{dY}{dt} = \frac{c}{b} \frac{1}{K} \frac{dK}{dt} + n_0 \alpha_2 \frac{a}{b} (1 - \frac{P}{P_0}) + \frac{\alpha\lambda}{b} (1 - \frac{T}{T_0}) \tag{9}$$

where

$$a = 1 + \mu_1 - \mu_2,$$
$$b = \alpha_1 + \alpha_2 - \alpha_2 \mu_2, \tag{10}$$
$$c = \alpha_1 - \alpha_2 \mu_1.$$

Finally from (2) and (8) the dynamic equation for $\beta = Y/K$ is given as:

$$\frac{1}{\beta} \frac{d\beta}{dt} \frac{a}{b} = -s\alpha_2 \frac{a}{b} \beta + \alpha_2 \frac{a}{b} s_0 + \alpha_2 \frac{a}{b} n_0(1 - \frac{P}{P_0}) + \frac{\alpha}{b} \lambda(1 - \frac{T}{T_0}) \tag{11}$$

If $\mu_1 = 0$, $\mu_2 = 0$, $P_0 \to \infty$, $T_0 \to \infty$, we get the same equation for β as obtained by Cigno (1981).

In the following, two cases are analyzed:

Case 1: $P_0 \to \infty$, $T_0 \to \infty$

$$\tag{12}$$

Case 2: $P_0 = $ const. , $T_0 = $ const.

3.2 Analysis

Case 1: P_0 and T_0 are infinite.

In this case the equation (11) for ß can be simplified to

$$\frac{1}{\text{ß}} \frac{d\text{ß}}{d\text{ß}} = -s\alpha_2 \frac{a}{\text{ß}} \text{ß} + \alpha_2 \frac{a}{b} s_0 + \alpha_2 \frac{a}{b} n_0 + \frac{\alpha\lambda}{b} . \tag{13}$$

The positive equilibrium point of (13) is given by:

$$\text{ß}^* = \frac{\alpha_2 a(s_0 + n_0) + \alpha\lambda}{s\alpha_2 a} > 0 \tag{14}$$

and which depends upon n_0 and λ.

To study the local and global stability of the equilibrium point (14) we proceed as follows:

By using the transformation

$$\text{ß}^* = \text{ß}^* + \text{ß}_1 ,$$

(13) may be written as

$$\frac{\text{ß}_1}{\text{ß}_1 + \text{ß}^*} = -s\alpha_2 \frac{a}{b} \text{ß}_1 \tag{15}$$

Linearlizing (15) we obtain the following condition for the stability of (14) as

$$\frac{a}{b} = \frac{1 + \mu_1 - \mu_2}{\alpha_1 + \alpha_2 - \alpha_2 \mu_2} > 0 \tag{16a}$$

when

$$0 < \mu_2 \leq 1 , \text{ i.e., for } a > 0 , b > 0 . \tag{16b}$$

The condition (16a) is automatically satisfied and ß tends to a finite limit ß^* . In the case when $\frac{a}{b} < 0$, ß would grow without limit and a typical condition for this is

$$\alpha_2 \mu_2 > 1 \text{ (i.e., } a > 0 , b < 0) . \tag{16c}$$

To obtain the global stability of the equilibrium point (14) we use the following Liapunov function,

$$V = \beta_1 - \beta^* \log(1 + \frac{\beta_1}{\beta^*}) , \tag{17}$$

and from (15), $\frac{dV}{dt}$ is obtained as

$$\frac{dV}{dt} = -s\alpha_2 \frac{a}{b} \beta_1^2 . \tag{18}$$

Application of Liapunov's Theorem shows that the condition for stability is the same as (16). It is therefore concluded that the equilibrium state is always stable under (16) and the economy would thus follow a sustainable path.

Case (2): P_0 and T_0 remain finite.

In this case the equations for P, T and β are given by (5), (6), and (11). The corresponding equilibrium point can be obtained as

$$P^* = P_0^*$$

$$T^* = T_0 \tag{19}$$

$$\beta^* = \frac{s_0}{s} \tag{19}$$

It is noted here that β^* in (19) is independent of n_0 and λ. Using the transformation,

$$P = P_1 + P^*$$
$$T = T_1 + T^*$$
$$\beta = \beta_1 + \beta^*$$

in (5), (6) and (11) we have,

$$\frac{dP_1/dt}{P_1 + P^*} = -\frac{n_0}{P_0} P_1 , \tag{20a}$$

$$\frac{dT_1/dt}{T_1 + T^*} = -\frac{\lambda}{T_0} T_1 , \tag{20b}$$

$$\frac{d\beta_1/dt}{\beta_1 + \beta^*} = -s\alpha_2 \frac{a}{b} \beta_1 - \alpha_2 \frac{n_0}{P_0} \frac{a}{b} P_1 - \frac{\alpha}{b} \frac{\lambda}{T_0} T_1 \ . \tag{20c}$$

As before, from (20), the condition for local stability is found to be the same as (16).

In order to study global stability of (20) consider the following Liapunov function

$$V = \beta_1 - \beta^* \log(1 + \frac{\beta_1}{\beta^*}) + k_1\{P_1 - P^* \log(1 + \frac{P_1}{P^*})\} + k_2\{T_1 - T^* \log(1 + \frac{T_1}{T^*})\} \ . \tag{21}$$

On using (20) we get from (21),

$$\frac{dV}{dt} = -\{s\alpha_2 \frac{a}{b} \beta_1^2 + k_1 \frac{n_0}{P_0} P_1^2 + k_2 \frac{\lambda}{T_0} T_1^2\} - \alpha_2 \frac{n_0}{P_0} \frac{a}{b} P_1 \beta_1 - \frac{\alpha}{b} \frac{\lambda}{T_0} T_1 \beta_1 \ . \tag{22}$$

It is seen from (22) that $\frac{dV}{dt}$ is negative definite provided (16) is satisfied. Thus, it is concluded that the equilibrium point (19) is globally asymptotically stable for $0 \leq \mu_2 \leq 1$.

This implies that the economy can follow a sustainable growth path under a wide range of economic parameters as compared to Solow (1956, 1974) and Stone (1980).

4 MODEL WITH LOGISTIC POPULATION AND TECHNOLOGY: EXHAUSTIBLE RESOURCE

When an exhaustible resource is explicitly considered as an input to the production process, then following Stiglitz (1974a,b), Cigno (1981, 1984a,b, this volume), Schuler (1979), the equation for the production function is modified as follows:

$$Y = K^{\alpha_1} L^{\alpha_2} R^{\alpha_3} T^{\alpha} \tag{23}$$

where $R(t)$ is the input of the resource at time t.

As pointed out earlier, in this case also, the labour population is generalized by taking into account the current resource stock $S(t)$, and the resource used, $R(t)$. The following form is assumed in the present analysis,

$$f = (\frac{P}{K})^{\mu_1} (\frac{Y}{P})^{\mu_2} (\frac{R}{S})^{\mu_3} \ , \mu_i > 0 \ . \tag{24}$$

The form (24) reduces to Stones (1980) case when $\mu_2 = 0$, $\mu_3 = 0$.

The equation governing the resource used is taken as Stiglitz (1974),

226

$$\frac{dS}{dt} = R .$$
(25)

The following equilibrium conditions are assumed in the analysis, Cigno (1981).

(i) $\quad p = \alpha_3 \frac{Y}{R}$,
(26)

(ii) $\quad \frac{dp/dt}{p} = \alpha_1 \frac{Y}{K}$,
(27)

where p is the price of the resource used at time t .

4.1 Dynamic Equation

In addition to $\beta = \frac{Y}{K}$ we study here the dynamic behaviour of $\gamma = \frac{R}{S}$, the rate of resource utilization (Stiglitz (1974a, b)) . Then from

$$\frac{d\gamma/dt}{\gamma} = \frac{dR/dt}{R} - \frac{dS/dt}{S}$$
(28)

and (25), (26) and (27) we get

$$\frac{d\gamma/dt}{\gamma} = \frac{dY/dt}{\gamma} - \alpha_1 \beta + \gamma$$
(29)

Keeping in view (23), (24) and proceeding as before, we have the dynamic equations for β and γ

$$\frac{d\beta/dt}{\beta} = -\frac{A + s\alpha_2 a}{b_1} \beta + \frac{\alpha_2 \mu_3}{b_1} \gamma + \frac{a_1 \alpha_2 s_0}{b_1} + \frac{a_2 \, an_0}{b_1}(1 - \frac{P}{P_0}) + \frac{\alpha\lambda}{b_1}(1 - \frac{T}{T_0}) \quad (30)$$

$$\frac{d\gamma/dt}{g} = -\frac{B}{b_1} \beta + \frac{b}{b_1} \gamma - \frac{c}{b_1} s_0 + \frac{a_2 \, an_0}{b_1}(1 - \frac{P}{P_0}) + \frac{\alpha\lambda}{b_1}(1 - \frac{T}{T_0}) , \quad (31)$$

where

$$a_1 = a - \mu_3 = 1 + \mu_1 - \mu_2 - \mu_3$$
(32a)

$$b_1 = b - \alpha_2 \mu_3 = \alpha_1 + \alpha_2 - \alpha_2 \mu_3 - \alpha_2 \mu_3 \qquad (32b)$$

$$A = \alpha_1 \alpha_3 + \alpha_1 \alpha_2 \mu_3 - s\alpha_2 \mu_3 \qquad (32c)$$

$$B = \alpha_1 (\alpha_3 + b) - cs \qquad (32d)$$

We again consider two cases for stability analysis under the conditions $a > 0$, $b > 0$, $B > 0$.
   (i) $P_0 \to \infty$, $T_0 \to \infty$.
   (ii) $P_0 = $ const., $T_0 = $ const. .

Case (i). In this case the dynamic equations for β and γ are given by:

$$\frac{d\beta/dt}{\beta} = -\frac{A + s\alpha_2 a}{b_1} \beta + \frac{\alpha_2 \mu_3}{b_1} \gamma + \frac{a_1 \alpha s_0 + A_1}{b_1}, \qquad (33)$$

$$\frac{d\gamma/dt}{\gamma} = -\frac{B}{b_1} \beta + \frac{b}{b_1} \gamma + \frac{A_1 - cs_0}{b_1}, \qquad (34)$$

where

$$A_1 = \alpha_2 a n_0 + \alpha\lambda > 0$$

since a is taken to be positive.
   The equilibrium point of the system (33) and (34) can now be obtained as:

$$\beta^* = \frac{-\alpha_2 s_0 (ba_1 + c\mu_3) - A_1 b_1}{\alpha_2 \mu_3 B - b(A + s\alpha_2 a)}$$

$$\qquad (35)$$

$$\gamma^* = \frac{A_1 (A + s\alpha_2 a) - A_1 B - cs_0 (A + s\alpha_2 a) - a_1 \alpha_2 s_0 B}{\alpha_2 \mu_3 B - b(A + s\alpha_2 a)}$$

which is positive if,

$a > 0, b > 0, a_1 < 0, b_1 < 0, B > 0, A_1 > 0,$

$A + s\alpha_2 a < 0, ba_1 + c\mu_3 < 0,$ (36a)

and

$A_1(A + s\alpha_2 a) - A_1 B - cs_0(A + s\alpha_2 a) - a_1 \alpha_2 s_0 B > 0$ (36b)

are satisfied.

Now by using the transformation,

$\beta = \beta^* + \beta_1$

$\gamma = \gamma^* + \gamma_1$

the equations (33) and (34) can be written as follows:

$$\frac{d\beta_1/dt}{\beta^* + \beta_1} = -\frac{A + s\alpha_2 a}{b_1} \beta_1 + \frac{\alpha_2\mu_3}{b_1}\gamma_1 \ ,$$ (37)

$$\frac{d\gamma_1/dt}{\gamma^* + \gamma_1} = -\frac{B}{b_1}\beta_1 + \frac{b}{b_1}\gamma_1 \ .$$ (38)

Using the Liapunov function,

$$V = \{\beta_1 - \beta^* \log(1 + \frac{\beta_1}{\beta^*})\} + k_3\{\gamma_1 - \gamma^* \log(1 + \frac{\gamma_1}{\gamma^*})\}$$

which after differentiating and using (38), gives

$$\frac{dV}{dt} = -\frac{A + s\alpha_2 a}{b_1}\beta_1^2 + k_3 \frac{b}{b_1}\gamma_1^2 + (\frac{a_2\mu_3 - k_3 B}{b_1})\beta_1 \gamma_1$$ (39)

Now choosing $k_3 = \frac{\alpha_2\mu_3}{B} = 0$ for $B > 0$ and using Liapunov's theorem, it is noted that $\frac{dV}{dt}$ is negative definite provided (36) is satisfied.

Similarly in Case (ii); i.e. when P_0 and T_0 are finite, the dynamics for β, γ, P, T are given by (30), (31), (5), and (6) .

The corresponding equilibrium points are given by

$$\beta^* = \frac{\alpha_2 s_0 (b a_1 + c \mu_3)}{b(A + s\alpha_2 a) - \alpha_2 \mu_3 B} \ ,$$

$$\gamma^* = \frac{c s_0 (A + s\alpha_2 a) + a_1 \alpha_2 s_0 B}{b(A + s\alpha_2 a) - \alpha_2 \mu_3 B} \ ,$$

(40)

$$P^* = P_0$$

$$T^* = T_0$$

which is positive under condition (36a) .

By using the transformation,

$$\beta = \beta^* + \beta_1 \ ,$$
$$\gamma = \gamma^* + \gamma_1 \ ,$$
$$P = P^* + P_1 \ ,$$
$$T = T^* + T_1 \ ,$$

equations (30), (31), (5) and (6) for β_1, v_1, P_1 and T_1 are transformed to the similar forms (37), (38), (20a) and (20b) respectively.

To study the stability of the equilibrium point (40), we use the following Liapunov function

$$V = \beta_1 - \beta^* \log(1 + \frac{\beta_1}{\beta^*}) + k_4 \{\gamma_1 - \gamma^* \log(1 + \frac{\gamma_1}{\gamma^*})\}$$

$$+ k_5 \{P_1 - P^* \log(1 + \frac{P_1}{P^*})\} + k_6 \{T_1 - T^* \log(1 + \frac{T_1}{T^*})\} \ ,$$

(41)

where k_4, k_5, k_6 are positive constants.

Proceeding as before and suitably choosing k_i (i = 4, 5, 6) we can see that $\frac{dV}{dt}$ is negative definite under condition (36a) implying that the logistic population and technology confer stability under less stringent conditions.

It is noted here that when $\mu_3 = 0$, the corresponding equilibrium state is unstable, as shown by Cigno (1981). Thus, from the above analysis, it is concluded that the otherwise unstable equilibrium state may become stable provided $\mu_3 \neq 0$ and (36) is satisfied. This implies that by a suitable choice of the labour-population relation involving rate of resource utilization term, the system may follow a sustainable growth path.

5 SUMMARY

In this paper, economic growth models with logistic population and technology have been investigated in the cases of plentiful resources (no scarce resource) and exhaustible resources. By a suitable choice of labour-population relation the dynamics of the out-put capital ratio and rate of resource utilization have been investigated.

In the case of a plentiful resource, it has been noted that the positive equilibrium point does not depend upon the intrinsic growth rate n_0 when population and technology follow logistic behaviour. It has been established that the economy may follow a sustainable path for a wide range of economic parameters (see condition (16)) . In the case of an exhaustible resource, it has been concluded that under appropriate conditions the economy which in other situations may be unstable, may follow a stable growth path.

6 REFERENCES

Arrow, K.J., (1962). The Economic Implications of Learning by Doing. Rev. of Eco. Studies, 29: 155-173.

Cigno, A., (1981). Growth with Exhaustible Resources and Endogeneous Populations. Rev. of Eco. Studies 43: 281-287.

_____, (1984a). Further Implications of Learning by Doing. Bull. of Eco. Res. 36: 97-108.

_____, (1984b). Consumption vs. Procreation in Economic Growth. In Economic Consequences of Population Change in Industrialised Countries. Edited by G. Steinmann, Springer-Verlag, Berlin.

Gompertz, B., 1825. On the Nature of the Function Expressive of the Law of Human Mortality, and on a New Method of Determining the Values of Life Contingencies. Philo. Trans. of Royal. Soc. 36: 513-585.

Kelley, A.C., 1974. The Role of Population in Models of Economic Growth. Amer. Eco. Rev. 64: 39-44.

Kosobud, R.F., and O'Neill, W.D., 1974. A Growth Model with Population Endogeneous. Amer. Eco. Rev. 44: 27-34.

LaSalle, J.P. and Lefschetz, S., 1961. Stability by Liapunov's Direct Method and Applications. Academic Press, New York.

Laitner, J., 1984. Resource Extraction Costs and Competitive Steady-State Growth. Int. Eco. Rev. 25: 297-324.

May, R.M., 1973. Stability and Complexity in Model Ecosystems. Princeton University Press, Princeton.

Phelps, E.S., 1966. Golden Rules for Economic Growth. John Wiley and Sons, New York.

Pielou, E.C., 1977. Mathematical Ecology. John Wiley and Sons, New York.

Schuler, R.E., 1979. The Long Run Limits to Growth: Renewable Resources, Endogeneous Population and Technical Change. Jour. Eco. Theory, 21: 166-189.

Solow, R.M., 1956. A Contribution to the Theory of Economic Growth. Quart. Jour. of Eco., 70: 65-84.

_____, 1974. Integenerational Equity and Exhaustible Resouces. Rev. Eco. Stud. with
      Symposium, 41: 29-45.
Stiglitz, J.E., 1974a. A Growth with Exhaustible Resources: Efficient and Optimal
      Growth Paths. Rev. of Eco. Stud. with Symposium, 123-138.
_____, 1974b. Growth with Exhaustible Natural Resource, the Competitive Economy.
      Rev. of Eco. Stud. with Symposium, 139-152.
Stone, R., 1980. A Simple Growth Model Tending to Stationarity. Eco. Jour. 90:
      593-598.
Suzuki, H., 1976. On the Possibility of Steadily Growing per Capita Consumption in an
      Economy with a Wasting and Non-Replenishable Resource. Rev. of Eco. Stud.,
      34: 527-535.
Swan, T.S., 1956. Economic Growth and Capital Accumulation. Eco. Record, 32:
      334-361.
Uzawa, H., 1961. Neutral Inventions and the Stability of Growth Equilibrium. Rev. of
      Eco. Stud. 28: 117-124.

A DYNAMIC PREDATOR-PREY MODEL FOR THE UTILIZATION OF FISHERY RESOURCES; A CASE OF TRAWLING IN LAKE KASUMIGAURA[1]

Y. KITABATAKE
Institute for Socio-Economic Planning, The University of Tsukuba, Sakura, Ibaraki 305, JAPAN.

1 INTRODUCTION

As D.W. Pearce (1976) has noted, the environment's main functions, from man's point of view, consist of "supplying "natural goods" such as beautiful landscape, supplying natural resources which are used to create economic goods, and supplying a "sink" into which the inevitable byproducts of economic activity can be discarded."

Lake Kasumigaura is unique in such a sense that all of the three functions coexist, though the first function is rather weak compared to the second and third function. The second function is composed of the utilization of lake water resources and of biological resources. Various land-based activities such as industry, households, and agriculture use water taken from the lake for their production or consumption activities, while fisherman engaging in aquaculture utilize a certain volume of water for nurturing their cultured fish. As a byproduct of this kind of water resource utilization processes, various kinds of pollutants and/or nutrients are discharged into the lake.

On the other hand, fishermen are also engaged in the utilization of fishery resources of the lake. Though this kind of utilization of biological system does not discharge any kind of pollutants and/or nutrients into the lake, their utilization processes do have a significant impact on the ecological structures of the lake and the impact may deteriorate the quality as well as quantity of lake environmental resources. For example, the decrease in the pelagic species population coupled with the increase in the demersal species population may deteriorate the lake water quality, for the latter species can feed on the bottom deposits and the nutrients are reinjected into the lake water by fishes' excretion process.

In this sense, the second and the third functions of the environment are closely related to each other. Ikeda and Yokoi (1980) analyzed the model dealing with the impact of nutrient enrichment in the Seto Inland Sea on the fishery resources. In their study, exogeneously given nutrient is assumed to affect the carrying capacity of plankton population and the death rate of small as well as large fish populations. The behaviour of each population level is described in differential equations and the stability, as well as the sensitivity with respect to nutrient enrichment of equilibrium

---

[1] Reproduced with slight modification, including the renaming of the title, from "Economic analysis of the utilization of fishery resources with predator-prey relationship: a case of trawling in Lake Kasumigaura," In: Economic Analyses of Man's Utilization of Environmental Resources in Aquatic Environments and National Park Regions, Edited by Y. Kitabatake, Research Report from the National Institute for Environmental Studies, Japan, No. 91, 1986, chapter 6.

points of differential equations are checked. Their model like most of fishery models rather simplifies man's behavioural characteristics such that catching of each fish species, for the unit time period, by fishermen is assumed to be some constant multiplied with the corresponding fish population, and to be independent of catching of the other species. Here we deal with a case where there is a decision making involved on the side of the fisherman to choose which species to catch.

Especially, we deal with the interactions between fisherman's behavioural characteristics represented by economic efficiency and ecological characteristics of the fishery resources, and clarify the economic rationale for the kind of operation pattern observed in the period between the middle of 1960's and the early part of 1980's in trawling in Lake Kasumigaura in which fishermen spend the greater part of their licensed trawling hours for the trawling of demersal species. This paper presents an extension of the author's previous work on the same subject (Kitabatake (1982)), in which the newly estimated production function for catching of demersal species is used in the predator-prey model.

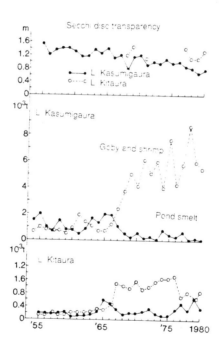

FIGURE 1 Trend of fish catch and water quality

2 A SUMMARY OF THE FISHERY MANAGEMENT OF LAKE KASUMIGAURA

The introduction of diesel-powered trawling in 1966, which substituted the traditional method of sailing trawling on most of Lake Kasumigaura and Lake Kitaura, had a significant impact on the interspecies relationships in Lake Kasumigaura. Figure 1 shows 1) the annual water quality variation in terms of Secchi depth in Lake Kasumigaura, 2) the annual variation of pelagic fish, represented by pond smelt, and 3) that of demersal fish, by goby and freshwater shrimp in each of two lakes.

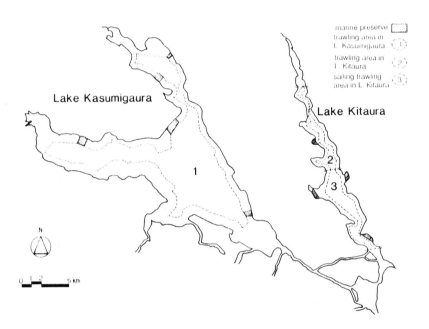

FIGURE 2. Licenced operation area in L. Kasumigaura and L. Kitaura

Figure 2 shows the operation areas of the lakes. The main fishing methods in use include fixed netting, trawling, and an indigenous fishing method of "Isaza-gorohiki ami" in which the net is hauled near the lake bottom toward an anchor for the distance of 400 to 1000 metres. In 1978, 74% of the total pond smelt catch, 43% of the total goby catch, and 26% of the total freshwater shrimp catch was taken by trawling.

Figures 3 and 4 show the biological aspects of fishery resources at Lake Kasumigaura. The seasonal variations of stomach contents of pond smelt is shown in Figure 3, whereas Figure 4 depicts the seasonal variations of population densities of

236

prey species. Furthermore, fishing by means of the trawl method is restricted to the period between 21 July and 31 December, during which time, as shown in Figure 3, pond smelt are observed to eat juvenile goby and freshwater shrimp larvae as well as opossum shrimp which has little market value. The above observations and research findings strongly suggest that there is a predator-prey relationship between the pelagic species of pond smelt and the demersal species of goby and freshwater shrimp and that the introduction of diesel-powered trawling on Lake Kasumigaura has had favorable impacts on the growth of demersal fish at the expense of the pelagic fish of pond smelt which has the higher market value.

Figure 5 and Table 1 show the economic characteristics of trawling in Lake Kasumigaura.

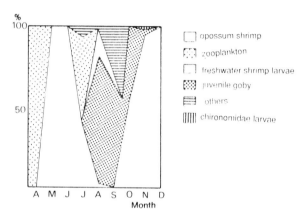

FIGURE 3 Seasonal variation of stomach content of pond smelt, Lake Kasumigaura

TABLE 1
Average catch by species for the period of four months - July to September, 1978

Species	Trawling in L. Kasumigaura	Trawling in L. Kitaura	Sailing Trawling in L. Kitaura
Pond Smelt (500 yen/kg)	328.3	1580.0	2531.3
Icefish (1200 yen/kg)	14.0	5.9	66.1
Goby (115 yen/kg)	2646.4	164.7	52.6
Freshwater shrimp (165 yen/kg)	549.3	48.8	89.5
Others	7.9	8.8	0

Source: Kitabatake (1981)

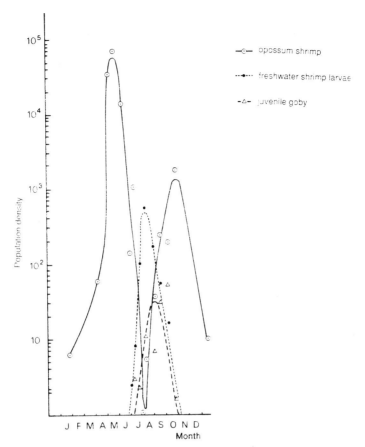

FIGURE 4 Seasonal variation of population densities of various prey species of smelt, Lake Kasumigaura

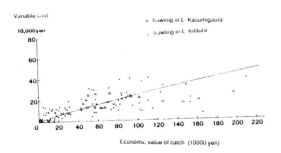

FIGURE 5 Costs of trawling (see text for explanation)

First, the estimated variable cost function of trawling is shown along with the sample data in Figure 5, where the estimation is based on the survey data (Kitabatake (1981, 1983)). The figure reveals that there is no significant difference in economic efficiency, in terms of variable cost per value of catch, of trawling in two lakes, where for those samples with high value of catch the data are not reliable due to the possibility of poaching. On the other hand, Table 1 summarizes the average catch by species for sampled fishery households. Contrary to the situation in Figure 5, there exists a large difference in the catch by species between trawling in Lake Kasumigaura and trawling as well as sailing trawling in Lake Kitaura. From these observations, we know that, although the cost effectiveness of trawling in two lakes are almost the same, economic viability of trawling in Lake Kasumigaura rests on demersal species, whereas that in Lake Kitaura on pelagic species of pond smelt.

3 THE MODEL OF FISHERY RESOURCES

3.1 The Biomass Equations without the Effects of Fishing

The following Volterra equations are postulated for the growth and population of pelagic and demersal species in lake water.

$$dX_1/dt = -a_1X_1 + b_1X_1X_2 \tag{1}$$

$$dX_2/dt = a_2X_2(1 - X_2/K_2) - b_2X_1X_2 \tag{2}$$

The assumptions underlying these equations are now summarized and are based on Smith (1974):

(1) populations can be adequately represented by a single variable, which ignores differences of age, sex, and genotype;

(2) the gross change in population can adequately be described by deterministic equations, which ignore random fluctuations in the environment with time;

(3) the effects of interactions between species are instantaneous;

(4) in the absence of pelagic species, the demersal species would follow the logistic equations, with intrinsic rate of increase a_2 and carrying capacity K_2; and

(5) the rate at which the demersal species are eaten is proportional to the product of the demersal and the pelagic populations.

The dynamics of the system, represented in equations (1) and (2) is shown in Figure 6. If the parameters a_1, a_2, b_1, b_2, and K_2 are all positive constants and fixed, and that a stationary state denoted by (X_1, X_2) in Figure 6 exists in the first quadrant, then it can be shown that the pelagic and demersal populations converge to the stationary state as t goes to infinity, where the stationary state is written as follows:

$$X_1 = a_2(1 - X_2/K_2)/b_2 \tag{3}$$

$$X_2 = a_1/b_1 \tag{4}$$

and the reader is referred to Leung and Wang (1976) for a proof.

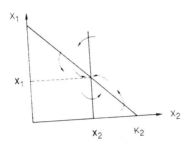

FIGURE 6 Phase plane diagram of pelagic and demersal species

3.2. The Biomass Equations Including the Effects of Fishing

The introduction of fishing activities modifies equations (1) and (2) as follows:

$$dX_1/dt = -a_1 X_1 + b_1 X_1 X_2 - n_1 Y_1 \tag{5}$$

$$dX_2/dt = a_2 X_2 (1 - X_2/K_2) - b_2 X_1 X_2 - n_2 Y_2 \tag{6}$$

where Y_1 and Y_2 are the catch rates of pelagic and demersal species, respectively, by an individual fishery household, and where n_1 and n_2 are the number of fishery households engaged in catching the pelagic and demersal species, respectively. The production functions are expressed as follows:

$$Y_1 = k_1 X_1^{\beta_1} U_1^{\alpha_1} \tag{7}$$

$$Y_2 = k_2 X_2^{\beta_2} U_2^{\alpha_2} \tag{8}$$

where k_1, k_2, α_1, and α_2, β_1, and β_2 are constants. A fishery household is assumed to own a diesel-powered boat to catch, during any time interval Δt, $Y_1 \Delta t$ of pelagic species and $Y_2 \, \Delta t$ of demersal species by expending U_1 and U_2 of labour force owned by the fishery household, respectively. Thus the amount of pelagic and that of all species caught by a fishery household are assumed in equation (8) to be a function of the stock of pelagic species and labour.

The behaviour of the system described by equations (5) through (8) is analyzed graphically in figures 7 and 8, for given values of $n_1, n_2, U_1,$ and U_2. The stationary points can be obtained by equating the right-hand side of each equation to zero. We can obtain three kinds of stationary point: (X_1^*, X_2^*) for $X_1^*, X_2^* > 0$, $(0, X_2^*)$ and $(0, 0)$.

Though, for each of these three kinds of stationary points, the stability analysis can be executed, we will execute it for the first kind only in the following. Figure 7 shows the alternative curves for $dX_1/dt = 0$ and $dX_2/dt = 0$ for the different values of parameters β_1 and β_2.

The phase diagrams are drawn in Figure 8.

In case of $0 < \beta_1 < 1$, there exists the possibility of multiple equilibria as shown in Figure 8c. The stability of a stationary point S follows from the negative angle at which the $dX_1/dt = 0$ and $dX_2/dt = 0$ lines intersect and is assured by checking that both eigenvalues of the linearized equations (5) and (6) at S have negative real parts (see Coddington and Levinson (1955)). The stability condition is written as follows.

$$a_2 X_2/K_2 + (\beta_1 - 1)n_1 k_1 X_1^{\beta_1 - 1} U_1^{\alpha_1} + (\beta_2 - 1) k_2 X_2^{\beta_2 - 1} U_2^{\alpha_2} > 0 \qquad (9)$$

which is clearly satisfied if $\beta_1 \geq 1$ and $\beta_2 \geq 1$.

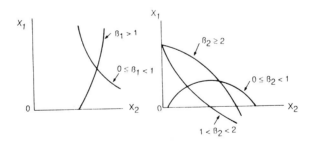

FIGURE 7 Isocline possibilities

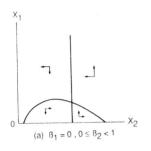 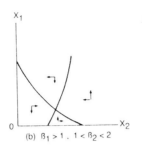 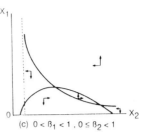

(a) $\beta_1 = 0 , 0 \leq \beta_2 < 1$ (b) $\beta_1 > 1 , 1 < \beta_2 < 2$ (c) $0 < \beta_1 < 1 , 0 \leq \beta_2 < 1$

FIGURE 8 Phase diagrams

In what circumstances will the extinction of pelagic or demersal species be prevented? It is clear from Figure 8 that the extinction of either species will not arise if the intersection of the curve of $dX_2/dt = 0$ and the horizontal axis gives the two distinct positive values of X_2, for the line of $dX_1/dt = 0$ assures $X_2^* > 0$, provided that the stability condition is satisfied. Especially, we know for an interesting case of $0 < \beta_2 < 1$, if the following conditions

$$(1 - \beta_2)K_2/(2 - \beta_2) > D$$
$$a_1/b_1 > D$$
$$\beta_1 \geq 1$$

where

$$D = (a_2/(K_2(1 - \beta_2)n_2k_2U_2^{\alpha_2}))^{1/(\beta_2 -2)} \tag{10}$$

are satisfied, there exists the stable solution with positive values.

4 MODELS OF FISHERY RESOURCE UTILIZATION

4.1 The model for a myoptic competitive fisherman

As to the individual rationality of fisherman, it seems quite natural to assume that "the individual fisherman attempts to maximize his net revenue flow at all times, based on the current conditions pertaining to stock abundance, prices, costs, regulations, and so on" (Clark (1985, p. 147)). Thus, in our case, the net revenue from the month's fishing are written as

$$\pi = P_1 Y_1 + P_2 Y_2 - w(U_1 + U_2) \tag{11}$$

where P_1 and P_2 are, respectively, the average price per unit of pelagic and demersal species, w the average wage rate per unit of labour, and Y_1 and Y_2 are the production function for pelagic species and for demersal species, respectively.

According to our hypothesis, for given values of X_1, X_2, P_1, P_2, and w, the individual fisherman maximizes π subject to the upper limit on the monthly labour hours,

$$0 \leq U_1 + U_2 \leq U \tag{12}$$

using monthly level of efforts (U_1^*, U_2^*).

4.2. A dynamic model for fishery management

For simplicity, assume that all fishermen have identical production functions, Y_1 and Y_2. Then the intraseasonal optimization model for a fishery sector is

$$\text{Maximize } \pi = \int_0^T (P_1 n_1 Y_1 + P_2 n_2 Y_2 - w(n_1 U_1 + n_2 U_2)) \exp(-rt) dt \tag{13}$$

subject to equations (5)-(8), (12), and to

$$X_i(t) \geq 0 \quad \text{for } 0 \leq t \leq T$$

$$X_1(0) = X_1 , X_2(0) = X_2$$

in which r is the relevant social rate of discount.

The optimal control model (13) can be solved by using the first-order conditions for optimizing the current-value Hamiltonian, H. H is written as:

$$H = d_1 X_1^{\beta_1} U_1^{\alpha_1} + d_2 X_2^{\beta_2} U_2^{\alpha_2} - d_3 U_1 - d_4 U_2$$

$$+ \phi_1(-a_1 X_1 + b_1 X_1 X_2 - n_1 k_1 X_1^{\beta_1} U_1^{\alpha_1})$$

$$+ \phi_2(a_2 X_2(1 - X_2/K_2) - b_2 X_1 X_2 - n_2 k_2 X_2^{\beta_2} U_2^{\alpha_2})$$

The first order conditions are written as

(1) $dx_i/dt = \partial H/\partial \phi_i$

(2) $d\phi_i/dt = r\phi_i - \partial H/\partial X_i$ for $i = 1,2$

(3) $U_1(t)$ and $U_2(t)$ maximize $H(X_1(t), X_2(t), \phi_1(t), \phi_2(t))$ subject to the constraints
$U_1(t) \geq 0$, $U_2(t) \geq 0$, $X_1(t) \geq 0$, and $X_2(t) \geq 0$.

(4) $\phi_i(T) \geq 0$, $\phi_i(T) X_i(T) = 0$, for $i = 1, 2$.

The sufficient condition (Arrow and Kurz (1970)) is not satisfied, for the integrand in model (13) and the right-hand sides of equations (5) and (6) are not concave in the variables X_1, X_2, U_1, U_2, take together. The right-hand side of equation (5) becomes concave if $\alpha_1, \beta_1 > 1$ and the concavity of the integrand requires $\alpha_1, \beta_1 < 1$, which are quite contradictory. Thus the optimal path of the problem is not discussed in this paper.

If a fishery household consistently maintains the constant labour units of U_1 and U_2, the biomass equations (5) and (6) will reach a stationary state where $dX_1/dt = 0$ and $dX_2/dt = 0$, provided that stability condition (9) is satisfied. Depending on the values of U_1 and U_2, therefore, there will be a sequence of stationary states. The stationary solution to the dynamic model (13) is derived by choosing the most profitable stationary state at which the long-run profit for a group of fishery households is maximized.

The stationary state (X_1^*, X_2^*) is derived from equations (5) through (8) such that

$$X_1^* = e_2(K_2 - X_2^*) - e_3 U_2^{\alpha_2} X_2^{*\beta_2 - 1} \tag{14a}$$

$$X_2^* = X_2 + e_4 U_1^{\alpha_1} X_1^{*\beta_1 - 1} \tag{14b}$$

where

$d_1 = P_1 n_1 k_1$, $d_2 = P_2 n_2 k_2$, $d_3 = n_1 w$, $d_4 = n_2 w$

$e_1 = K_2 X_1/(K_2 - X_2)$, $e_2 = X_1/(K_2 - X_2)$

$e_3 = X_1 K_2 n_2 k_2/(a_2(K_2 - X_2))$, $e_4 = X_2 n_1 k_1/a_1$

By substituting X_1^* in (14b) into (14a), equations (14a) and (14b) are reduced to the equation $F(X_2^*) = 0$. The root (X_2^*) of the equation $F(X_2^*) = 0$ is, in practice, obtained by such an approximation method as the method of successive approximation (McCracken and Dorn (1966)).

The substitution of the stationary state (X_1^*, X_2^*) into the integrand on the right-hand side of model (13) leads to the following form for the long-run optimization problem for a group of fishery households:

$$\text{maximize } \pi = d_1 X_1^{*\beta_1} U_1^{\alpha_1} + d_2 X_2^{*\beta_2} U_2^{\alpha_2} - d_3 U_1 - d_4 U_2 \tag{15}$$

subject to equation (12) where X_1^* and X_2^* are functions of U_1 and U_2 as specified in (14).

The sufficiency condition for the optimum is satisfied if the long-run profit function in problem (15) is a concave function of U_1 and U_2. Unfortunately, the sufficiency condition is not satisfied. Thus the optimum solution to problem (15) has to be found by numerical analyses.

5 EMPIRICAL ESTIMATES

Catch and production-effort data for empirical estimations of equation (7) are taken from the survey questionnaire conducted by the National Institute for Environmental Studies (NIES) (Kitabatake and Aoki (1980)), the survey period of which was four consecutive months, June to September, of 1978, and in which the questionnaires were sent for a self-reporting survey to all of the fishery households engaged in fixed netting, trawling, Isaza-gorohiki ami, and carp culture operations. In this study, a part of the NIES survey data is utilized, which is related to trawl netting. The pond smelt population for each of three consecutive months of 1978, over which catch and effort data is available for trawling, is estimated on catch and effort data taken from source data of the Ibaraki Statistics and Information Office (1980), and on observations made by Hamada (1978) that nearly 90-99% of the pond smelt population is estimated to be taken by fishermen every year. Table 2 shows the estimated monthly population of pond smelt.

TABLE 2

Estimated monthly populations of pond smelt.

	Lake Kasumigaura			Lake Kitaura (trawling area)		
	July	August	September	July	August	September
Catch per fishing days (kg/day)	6.68	1.46	0.71	59.85	27.39	12.30
Estimated population (kg)	52662	39760	8689	101226	60837	27837

Source: Kitabatake (1982)

On the other hand, catch and production effort data for empirical estimation of equation (8) are taken from the survey questionnaire conducted by NIES (Kitabatake, Kasuga, and Onuma (1984)), the survey period of which was April 1981 to March 1982, and in which ten experienced fishermen, employing the specific fishing gears of Isaza-gorohiki ami and trawling, were asked to write down the daily results of fishing operation. The items included in the survey were: (1) the type of fishing gears employed; (2) the fishing efforts involved, such as fishing hours per day; (3) the species composition of the catch; (4) the area of trawling. As to the monthly variation of demersal fish population, we refer to the study done by Onuma, Takahashi, Suzuki and Huzitomi (1984). Figure 9 shows the spatial distribution of seven sampling points in Lake Kasumigaura where they collected by the use of a towning net 112 samples during the period from April 1981 to April 1982.

FIGURE 9 Spatial distribution of sampling points in Lake Kasumigaura

Then they investigated the species composition, standing crop and prey organism composition in the stomachs of the benthic fishes. From their investigation, we first know that "Chichibu" (goby) and "Tenagaebi" (freshwater shrimp) are two dominant demersal species. Secondly, from the seasonal variation of the carapace length distribution of the prawn for "Tenagaebi" and the seasonal variation of body length distribution of the Chichibu, we know that the significant change in age distribution occurs in both species during summer season.

The estimated production function for pond smelt is as follows:

$$\ln Y_1 = -9.318 + 0.696 \ln U_1 + 0.610 \ln F_1 + 1.096 \ln X_1$$
$$\quad\quad\quad\quad (3.47) \quad\quad\quad (2.18) \quad\quad\quad (4.68)$$

$$R^2 = 0.5327, \ N = 57 \tag{16}$$

where the t-values are given in brackets, and

Y_1 is the monthly catch
U_1 is the monthly fishing labour hour which is the number of persons engaged in operations of capture multiplied by trawling hours,
F_1 is the monthly consumption of diesel oil in terms of the number of fuel cans (200 litres).

Equation (16) is estimated on data taken from Lake Kasumigaura and Lake Kitaura, provided that the difference in topography of fishery grounds between the two sublakes is assumed to have no effects on the technological efficiency in the catch of pond smelt. In the regression analysis, only those samples have been chosen whose catch contain more than 80% in terms of wet weight basis of pond smelt for the estimation of equation (16).

Table 3 compiles the data for estimating the production function for demersal species, where the sample data were excluded, in which some survey items such as the trawling area were unanswered. Sample points listed in the sixth column of Table 3 corresponds to those points in Figure 9. From Table 3, the production function for demersal species is estimated as follows:

$$\text{Log } Y_2 = -2.3678 - 0.07909 \text{Log } Z + 0.63839 \text{ Log } X_2 + 1.1765 \text{ Log } U_2$$
$$\quad\quad\quad\quad\quad (-1.129) \quad\quad\quad (2.841) \quad\quad\quad\quad (6.166)$$

$$R^2 = 0.7682, \ N = 20 \tag{17}$$

where

Y$_2$ is the catch of demersal species over 20 days period in July to October

Z is the catch of pond smelt over the same period

U$_2$ is the fishing labour hour over the same period

X$_2$ is the estimated demersal species population over the same period.

In the regression analysis, only those samples have been chosen whose total catch contain more than 80% of goby and freshwater shrimp.

Substituting the sample mean data of F$_1$ and Z into equations (16) and (17) and multiplying U$_2$ by 1.5 so as to transform the length of the period into a month, the following production functions are obtained:

$$Y_1 = 1.471 \times 10^{-4} U_1^{0.696} X_1^{1.096} \tag{18}$$

$$Y_2 = 0.00604 U_2^{1.1765} X_2^{0.63839} \tag{19}$$

where U$_i$ are monthly labour hours.

6   WHY DO GOOD OLD DAYS HARDLY RETURN?

6.1. The case of a myopic competitive fishermen

For the maximization of net revenue equation (11) subject to the constraint (12), price and cost data must be specified. The price data, P$_1$ = 500 yen per kg and P$_2$ = 140 yen per kg , are estimated by interviewing the fishermen. The wage datum, w = 653 yen per labour hour, is estimated from the Ministry of Agriculture, Forestry and Fisheries (1980b).

Figures 10 and 11 summarize the computation results, in which the total profit curve, the profit curve of catching pelagic specy, and the profit curve of catching demersal species are drawn as the function of U$_1$ + U$_2$, that of U$_1$, and that of U$_2$, respectively.

TABLE 3

Data for the estimation of production function of catching demersal species

No. of fishermen	time period	Catch of demersal species (kg)	Catch of pelagic species (kg)	Fishing hour (hr)	Sampling points related to trawling area	Estimated demersal species population (kg)
1	7/21-8/10	423.2	10.8	34	B - G	128793.6
	8/11-8/31	94.0	7.0	3	F, G	456647.1
	9/1-9/20	768.8	4.5	15	F, G	776534.9
	9/21-10/10	2358.8	20.1	63	F, G	406253.3
2	7/21-8/10	578.7	18.3	25	B - G	128793.6
	8/11-8/3	1890.2	0.1	33	B - E	370667.3
	9/1-9/20	2291.1	10.1	37	A - G	617471.4
	9/21-10/10	2982.2	2.7	42	B - G	573366.3
3	7/21-8/10	713.0	80.0	22	B - G	128793.6
	8/11-8/31	311.0	11.5	11	B - G	399315.5
	9/1-9/20	2203.0	16.0	29	B - G	657350.5
	9/21-10/10	2212.0	31.8	34	B - G	573366.3
4	7/21-8/10	400.5	18.2	27	B - E	404911.7
	8/11-8/31	1234.5	5.5	27	B - E	370667.3
	9/1-9/20	2568.0	0.1	27	B - E	597783.4
	9/21-10/10	4762.4	14.3	30	B - E	656931.2
5	7/21-8/10	407.9	44.2	30	A - G	128793.6
	8/11-8/31	149.0	11.2	11	A - G	399310.5
	9/1-9/20	1489.5	8.7	37	A - G	657350.5
	9/21-10/10	1262.2	22.4	53	A - G	573366.3

Source: Kitabatake (1986)

In both figures, the current restriction of monthly operation time $(U = 60$ hours) is assumed. The difference in two figures rests on the assumed values of X_1 and X_2 in equations (17) and (18). Figure 10 corresponds to the case of $(X_1, X_2) = (101226, 128793)$, where the value of X_1 is the estimated pond smelt population in July in Lake Kitaura (see table 2) and that of X_2 the smallest value of estimated demersal species population in table 3. Figure 11 corresponds to the case of $(X_1, X_2) = (52662, 657350)$, where the value of X_1 is the estimated pond smelt population in July in Lake Kasumigaura (see table 2) and that of X_2 the largest value of estimated demersal species population in table 3. Thus figure 10 represents the case of relative abundancy of pond smelt population such as that of Lake Kitaura, while figure 11 represents the case of relative abundancy of demersal species population such as that of Lake Kasumigaura.

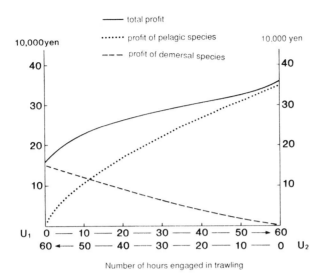

FIGURE 10 Profit Curves with $(X_1, X_2) = (101226, 128793)$

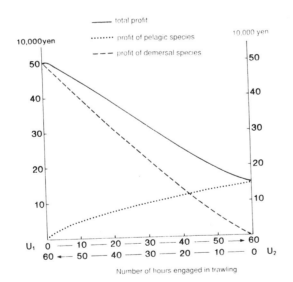

FIGURE 11 Profit Curves with $(X_1, X_2) = (52662, 657350)$

The optimal operation patterns for the individual fisherman are calculated as follows:

For the fisherman in Lake Kitaura,

$U_1 = 60$ hours, $U_2 = 0$
$Y_1 = 778,04$ kg , $Y_2 = 0$
$\pi = 349,842$ yen .

For the fisherman in Lake Kasumigaura,

$U_1 = 1$ hour, $U_2 = 59$ hours,
$Y_1 = 22.0$ kg , $Y_2 = 3788.5$ kg
$\pi = 502,207$ yen.

These computation results correspond rather well to the survey data listed in Table 1 and Table 2, and imply the two things: 1) the kind of operation pattern as well as the dominance of pelagic species over demesal species, which existed in Lake Kasumigaura under the fishing method of sailing trawling, is hardly realized as long as a myopic competitive fisherman employ diesel- powered trawling, even if the restriction of operation time is imposed; 2) when we consider the difference in reproduction rate between pelagic specy of pond smelt and demersal species, the optimal operation pattern for Lake Kitaura is to decrease the pelagic species population relative to the demersal species population, eventually. Thus, the reason why in Lake Kitaura the relative abundancy of pond smelt population has been observed may lie in that trawling area is restricted to the northern half of the lake and the topography of the lake is more complex than that of Lake Kasumigaura as shown in Figure 2. In this sense, the results of public decision to open the whole part of Lake Kitaura to diesel-powered trawling in 1983 would be quite interesting.

6.2. The case of long-run profit maximization

In this section, the long-run profit optimization problem for a group of fishermen, which is specified in equations (14) and (15), is calculated numerically for the case of Lake Kasumigaura.

Before specifying the parameter values, substitution of equations (18) and (19) into the differential equations (5) and (6) needs a couple of comments. First, equations (18) and (19) were estimated in terms of monthly or twenty days period data from July to September or to October. Since it has been assumed that there are continuous fishing activities, differential equations (5) and (6) should be considered to simulate with the unit time of a month (720 hours) a sequence of three months, July to September, for subsequent years. Second, labour hours U_1 and U_2 in equations (18) and (19) are transformed into labour units u_1 and u_2 by dividing by the unit time of 720 hours:

$$Y_1 = 0.0143 u_1^{0.696} X_1^{1.096} \tag{20}$$

$$Y_2 = 13.89486 u_2^{1.1765} X_2^{0.63839} \tag{21}$$

For numerical computation of the long-run optimization problem, the thirteen constants of the system $n_1, n_2, \alpha_1, \alpha_2, \beta_2, k_1, k_2, \beta_1, a_1, a_2, X_1, X_2$, and K_2 should be specified. Six of them, $\alpha_1, \beta_1, \alpha_2, \beta_2, k_1, k_2$ are estimated in equations (20) and (21), whereas n_1 and n_2 are estimated on data taken by the Ministry of Agriculture, Forestry and Fisheries (1980a) such that $n_1 = n_2 = 295$. It is assumed that $X_1 = 1125800 kg$, $X_2 = 1384000$ kg, and $K_2 = 12378000$ kg, which were derived on the basis of the following two assumptions. First, the stationary state (X_1, X_2) of pelagic and demersal fish populations is assumed to be approximated by the state that existed before 1965 (prior to the introduction of diesel-powered trawling). Specifically, X_1 is assumed to be equal to the 10 year (1956 to 1965) average of the pond smelt catch divided by man's exploitation rate of 95%, and X_2 to equal the ten year average of the goby and freshwater shrimp catch divided by man's exploitation rate of 70%. Second, the carrying capacity, K_2, of demersal species is assumed to be the goby and freshwater shrimp catch of 1978 divided by man's exploitation rate of 70%. In the above estimates, the assumed exploitation rates are based on Hamada (1978). The constant a_1 was assumed to be given by $a_1 = 0.167$, a figure based on the report by Kasebayashi and Nakano (1961) that the number of two-year-old fish is very small, on average being only 0.02-0.4% of the total number of fish. Since differential equations (5) and (6) are assumed to simulate the sequence of three months, July to September, a_1 is estimated to be 1/6. The a_2 is calculated in the following way. First we transformed the variables X_1, X_2, K_2, X_1, X_2, by dividing by $B_1 = 1125800$ kg, into x_1, x_2, k_2, x_1, x_2, respectively. Then the stationary relationship of equation (4) enables us to calculate

$$b_1 = a_1/x_2$$

The value of b_2 is calculated, based on the estimated relationship derived from the data ranging from the year 1956 through 1979 and plotted in figure 2 of $X_1 = 1287.6-0.1691 X_2$, where the regression coefficient is utilized to assume the relation $b_1/b_2 = 0.1691$. Finally, the a_2 is calculated to be 0.903 in terms of the stationary relationship of equation (3).

The optimum solution, which satisfies the stability condition of equation (9), to the long-run optimization problem (15) is obtained as follows,

$$U_1^* = 14.4 \text{ hours}, \qquad U_2^* = 38.16 \text{ hours}, \qquad \pi^* = 380.3 \text{ million yen},$$

$$X_1^* = 1.89 \text{ t}, \ X_2^* = 6133.55 \text{ t}.$$

Since the sum of U_1^* and U_2^* is less than the current restriction, 60 hours, on operation time, this solution is also valid for the case of restricted operation time.

Thus, even if we assume the kind of group rationality represented by the long-run profit maximization problem, the above numerical simulation clearly shows that the good old days hardly return in Lake Kasumigaura, provided that the dominant factors explaining the level and composition of fishery resources in Lake Kasumigaura are the predator-prey relationships and fishermen's behaviours in trawling, which our postulated model tried to simulate.

7 CONCLUDING REMARKS

In this paper, the model of fishery resources with a predator-prey relationship and the models of man's utilization of fishery resources for the period of three months in summer are constructed for the case of Lake Kasumigaura, and which are based on observational data and research findings. As to the latter models simulating the behavioural characteristics of fishermen, two models are used, which are the model for a myopic competitive fisherman, and the model for a long-run profit maximizing fisherman. The individual production function of pelagic species (pond smelt) and of demesal spieces (Chichibu and Tenagaebi) are estimated, for the fishing method of diesel-powered trawling.

Both the myopic model and the long-run profit maximization model insist that the kind of operation pattern as well as the dominance of pelagic species over demersal species, which existed in Lake Kasumigaura under the fishing method of sailing trawling, is hardly realized as long as fishermen employ diesel-powered trawling. This is mainly due to the estimated form of production functions in which both the production function of pelagic specy and that of demersal species depend on the respective population level. This stock dependency of production efficiency in trawling operation forces to bend down the curves of X_1 against X_2 in figures 6-8 for which the growth rate of demersal species population becomes zero, and which forces the stationary population level of demersal species in the biomass equations including the effects of fishing to be greater than that in the biomass equations without the effects of fishing as shown in equation (12).

Thus, even if the current biomass structures of Lake Kasumigaura is known to be unfavorable for preventing the eutrophication of lake, we have to keep in mind in divising management measures that the existing biomass structures reflect the individual fisherman's economically rational response to given technological, economic, and ecological conditions pertaining to fishery resources in Lake Kasumigaura.

8 ACKNOWLEDGEMENTS

This paper reports one of the works executed in a research group on the structural situations and the functional role of fish in the ecosystem of Lake Kasumigaura, in the special research project of the National Institute for Environmental Studies, Japan, concerning the eutrophication of Lake Kasumigaura. The research group consisted of scientists from three different institutions, The National Institute for Environmental Studies, The Freshwater Fisheries Experiment Station of Ibaraki Prefecture, and the Ocean Research Institute, University of Tokyo. This paper would not have been written without the valuable experiences obtained in this research group, and, especially, without the fishermen's cooperation in our research activities. Thus the author expresses his sincere thanks to the fishery cooperative in Lake Kasumigaura and to the members of the group, especially, to Dr. S. Kasuga, Messers J. Takahashi, Y. Onuma and K. Tachikawa, and Professor S. Tanaka. The author also wishes to express his thanks to Professor Jiro Kondo, former Director of NIES, who initiated my interest in the integration of ecological and economic models. The usual dislaimers apply.

9 REFERENCES

Arrow, K.J., M. Kurz (1970). Public Investment, the Rate of Return, and Optimal Fiscal Policy, Baltimore: Johns Hopkins University Press.
Coddington, E., N. Levinson (1955). Theory of Ordinary Differential Equations, New York: McGraw-Hill.
Hamada, A. (1978). "Fishes of Lake Kasumigaura" in "The impact of human activities on ecosystem dynamics of Lake Kasumigaura and its basin area" Research Report B3-R12-1, Special Research Project on Environmental Science supported by grants in aid for scientific research, Ministry of Education, Culture, and Science. Tokyo: University of Tokyo, pp. 143-150 (in Japanese).
Ibaraki Statistics and Information Office (1980). 1978-1979 Annual Statistics of Agriculture, Forestry and Fisheries in Ibaraki, Ministry of Agriculture, Forestry and Fisheries, Mito: Ibaraki Norin Tokei Kyokai. (in Japanese).
Ikeda, S. and T. Yokoi (1980). "Fish population dynamics under nutrient enrichment - a case of the East Seto Inland Sea," Ecological Modeling, 10: 141-165.
Kasebayashi, T., I. Nakano (1961). "Fishery biological studies of pond smelt, hypomesus olidus, in Lake Kasumigaura," Research Report 6, Tsuchiura: Kasumigaura-Kitaura Fisheries Office of Ibaraki Prefecture. (in Japanese).

254

Kitabatake, Y. (1981). "Water pollution effects on a fishing method of trawling at Lake Kasumigaura," Res. Rep. Natl. Inst. Environ. Stud. Jpn., No. 24, 65-80. (in Japanese).

_____ (1982). "A dynamic predator-prey model for fishery resources: a case of Lake Kasumigaura," Environ. Plann. A. 14: 225-235.

_____ (1983). "Economic analysis of trawling in Lake Kasumigaura," J. of the North Japan Fisheries Economics, No. 13: 66-75. (in Japanese).

_____, S. Kasuga, and Y. Onuma (1984). "Survey analysis of Isaza-gorohiki ami and trawling in Lake Kasumigaura," Res. Rep. Natl. Inst. for Environ. Stud. Jpn., No. 53: 21-28.

Kitabatake, Y., and Y. Aoki (1981). "Empirical study of the eutrophication effects on commercial fishing at Lake Kasumigaura," Res. Rep. Natl. Inst. Environ. Stud. Jpn., No. 24: 27-51(in Japanese).

Leung, A. and A-Y Wang (1976). "Analysis of methods for commercial fishing: mathematical and economical aspects," Econometrica 44(2), 295-303.

Ministry of Agriculture, Forestry and Fisheries, (1980a). 1978 Annual Report of the Fisheries and Culture Fisheries Production Statistics, Tokyo: Norin Tokei Kyokai. (in Japanese).

Ministry of Agriculture, Forestry and Fisheries (1980b): 1978 Annual Report of the Fisheries Economy Survey (Fishermen's Households), Tokyo: Norin Tokei Kyokai. (in Japanese).

Onuma, Y., J. Takahashi, K. Suzuki and M. Huzitomi (1984). "Study on the production of benthic animals in Lake Kasumigaura-1: biomass and feeding habit of gobiidae and the prawn," Natl. Inst. for Environ. Stud. Jpn., No. 53: 61-84.

Pearce, D.W. (1976). Environmental Economics, London: Longman.

Smith, J.M. (1974). Models in Ecology. Cambridge: Cambridge University Press.

Suzuki, K., and T. Ida (1977). "Study on the productive structures of fishery resources in Lake Kasumigaura-I," Research Report 14, Tamatsukuri: Ibaraki Freshwater Fisheries Experimental Station, 1-19. (in Japanese).